Optimal Data Analysis

Optimal Data Analysis
A Guidebook With Software for Windows

Paul R. Yarnold and Robert C. Soltysik

American Psychological Association
Washington, DC

Copyright © 2005 by the American Psychological Association. All rights reserved. Except as permitted under the United States Copyright Act of 1976, no part of this publication may be reproduced or distributed in any form or by any means, or stored in a database or retrieval system, without the prior written permission of the publisher.

Published by
American Psychological Association
750 First Street, NE
Washington, DC 20002
www.apa.org

To order
APA Order Department
P.O. Box 92984
Washington, DC 20090-2984
Tel: (800) 374-2721; Direct: (202) 336-5510
Fax: (202) 336-5502; TDD/TTY: (202) 336-6123
Online: www.apa.org/books/
E-mail: order@apa.org

In the U.K., Europe, Africa, and the Middle East, copies may be ordered from
American Psychological Association
3 Henrietta Street
Covent Garden, London
WC2E 8LU England

Typeset in Garamond by World Composition Services, Inc., Sterling, VA

Printer: Automated Graphics Systems, White Plains, MD
Cover Designer: Aqueous Studio, Arlington, VA
Technical/Production Editor: Dan Brachtesende

The opinions and statements published are the responsibility of the authors, and such opinions and statements do not necessarily represent the policies of the American Psychological Association.

Library of Congress Cataloging-in-Publication Data
Yarnold, Paul R.
　Optimal data analysis : a guidebook with software for windows / Paul R. Yarnold and Robert C. Soltysik.—1st ed.
　　p. cm.
　Includes bibliographical references and index.
　ISBN 1-55798-981-8
　1. Optimal designs (Statistics) I. Soltysik, Robert C. II. Title.
　QA279.Y367 2004
　003—dc22
                                                        2004010917

British Library Cataloguing-in-Publication Data
A CIP record is available from the British Library.

Printed in the United States of America
First Edition

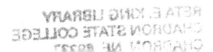

*For my wife, Loretta J. Stalans;
my daughter, Maggie S. Yarnold;
my mother, Helen M. Yarnold;
my father, James. K. Yarnold;
and my siblings, teachers, and friends.
—Paul R. Yarnold*

*For Samuel
—Robert C. Soltysik*

Contents

Preface xi
Acknowledgments xiii

Chapter 1   Introduction to the ODA Paradigm 3
            What Is ODA? 3
            Why Is ODA Superior to Other Data Analysis Programs? 5
            Who Is the Audience for This Book and Software? 6
            How Should the Reader Use This Book? 6
            Basic Steps and Key Concepts 7
            Historical Perspective 10
            Thirty Hypothetical Applications 11
            Now It Is Time to Start Analyzing Data 27

Chapter 2   Using the ODA Software 29
            The ODA Commands 29
            Running the ODA Software 37
            Using ODA With PFE 38
            Creating a Data Set for Analysis by ODA 45

Chapter 3   Evaluating Classification Performance 57
            How to Obtain an ODA Model 61
            Selecting Among Multiple Optimal Models 64
            Assessing Model Stability 68
            Standardizing Transformations 70

Chapter 4   Evaluating Statistical Significance 73
            Analytic Methodology 74
            Fisher's Randomization Methodology 77

Monte Carlo Methodology 78
Specifying the Type I Error Rate 80
A Priori Alpha Splitting 83

Chapter 5 Two-Category Class Variables 87
Applications Involving Binary Attributes 87
Applications Involving Polychotomous Attributes 91
Applications Involving Ordinal Attributes 93
Applications Involving Continuous Attributes 101

Chapter 6 Multicategory Class Variables 107
Applications Involving Binary Attributes 108
Applications Involving Polychotomous Attributes 108
Applications Involving Ordinal Attributes 112
Applications Involving Continuous Attributes 115

Chapter 7 Reliability Analysis 121
Inter-Rater Reliability 122
Parallel Forms Reliability 128
Split-Half Reliability 130
Temporal Reliability 132
Nonlinear Reliability 135
Intraclass Correlation 138

Chapter 8 Validity Analysis 141
Hold-Out (Cross-Generalizability) Validity 142
Construct Validity 148
Convergent and Discriminant Validity 149

Chapter 9 Optimizing Suboptimal Multivariable Models 155
Optimizing Fisher's Linear Discriminant Analysis 157
Optimizing Logistic Regression Analysis 160
Optimizing Complex Models 165

Chapter 10 Multiple Sample Analysis 167
Pooling Samples and Simpson's Paradox 168
The ODA Generalizability Algorithm 170
Evaluating Model Generalizability Across Samples 172

Analyzing Randomized Block Designs   178
Optimizing Multiple Suboptimal Multiattribute Models   181

Chapter 11   Sequential Analyses   187
Identifying Structure in Markov Transition Tables   187
Analyzing Turnover Tables   193
Autocorrelation (Time Series) Analysis   198
Repeated Measures (Within-Subjects) Analysis   203
Single-Case (N-of-1) Analysis   207

Chapter 12   Iterative Decomposition Analysis   209
Stopping Rules for Iterative Analyses   212
Structural Decomposition With Sequential Data   214
Reliability, Bias, and Random Error   223
Validity, Bias, and Random Error   225

Epilogue   The Future of ODA   229
General-Purpose MultiODA Models   232
Special-Purpose MultiODA Models   233
Nonlinear Classification Tree Analysis   237
Users of ODA   239

Appendix A: Dunn and Sidak Adjusted Per-Comparison p 241

Appendix B: Troubleshooting: Common Problems and Their Possible Solutions   249

References   253
Index   275
About the Authors   287

Preface

Surveying the battlefield, Aristotle asked, "Why does the cannon ball choose its path to fly?" (see Boring, 1950). This question motivated philosophical study of will, destiny, and so forth, but did not do much to help the accuracy of the gunners who were firing the cannon. Newton thought the question silly ("the cannon ball chooses its path to fly because it wants to!") and felt that the more interesting question was "why does the path the cannon ball chooses to fly *change*?" The notion of change led to the development of the derivative, calculus, the hypotheticodeductive method, and the idea of control: the genesis of the first renaissance. Fisher rephrased the question: "Why does the path the cannon ball chooses to fly change *in a variable manner* from one shot to another?" Awareness of variability introduced the world to the concept of random error and triggered the second renaissance. The optimal data analysis (ODA) paradigm rephrases the question yet again: "How can we hit the target with the cannonball?"

In order to determine whether ODA is the appropriate method of analysis for any particular data set, it is sufficient to consider the following question: When you make a prediction, would you rather be correct or incorrect? If your answer is "correct," then ODA is the appropriate analytic methodology—by definition. That is because, for any given data set, ODA explicitly obtains a statistical model that yields the theoretical maximum possible level of predictive accuracy (e.g., number of correct predictions) when it is applied to those data. That is the motivation for ODA; that is its purpose. Of course, it is a matter of personal preference whether one desires to make accurate predictions. In contrast, alternative non-ODA statistical models do not explicitly yield theoretical maximum predictive accuracy. Although they sometimes may, it is not guaranteed as it is for ODA models. It is for this reason that we refer to non-ODA models as being *suboptimal*.

Development of this book and software system truly reflected a collaborative scientific research project and required twenty years of our mutual effort to accomplish. We began conducting collaborative research concerning the use of ODA optimization methodologies in the context of the statistical classification problem in 1977, and we jointly discovered the

ODA statistical paradigm between 1987 and 1991. Although many discoveries occurring during this time span have not yet been published, many have been elaborated in a series of ten invited lectures presented at the 1992 and 1993 joint national meetings of the Institute of Management Sciences and the Operations Research Society of America. In 1992, Rob began to program the software while being largely unaware of the applications for which it would be used and, simultaneously, Paul began to develop the applications and write the book while being largely unaware of the software with which data would be analyzed. The program and the book converged in only three iterations! We wish to make clear that we contributed equally to the development of the ODA paradigm and to this book and software system.

This book represents the only comprehensive exposition of the ODA statistical paradigm yet to be published. The program and book were written to provide people who have never heard of the ODA paradigm with sufficient information to be able to conduct a variety of ODA analyses quickly, with a minimum of effort. To take maximum advantage of the software, we recommend that you carefully and sequentially read each chapter. Every example problem should be solved as it is encountered during this reading. Run different variations of each example problem (e.g., with vs. without weighting by prior odds; with vs. without a directional alternative hypothesis, etc.), to become more familiar with the effect of software parameter settings on the outcome of the analysis. Of course, if possible, you should attempt to replicate the type of analysis illustrated in each example problem using personal data sets and publish the results.

You will find further information about the ODA model at www.apa.org/books/resources/yarnoldsoltysik. We've developed this Web site to provide you with a list of current publications and upcoming workshops on ODA, a venue for commenting on the software, and frequently asked questions that will evolve with your comments.

Acknowledgments

This project could not have been completed without support provided to us by many of our academic colleagues and friends. In particular, we express our great appreciation to the following people, listed in no particular order: Barbara M. Yarnold, Loretta J. Stalans, Kevin P. Weinfurt, Fred B. Bryant, Jamshid Hosseini, Gary Koehler, Leonard Kent, Arshad Khan, Al Levin, Mike Levine, Bill O'Neill, Steve Roy, Herb Stenson, Conrad Stille, Harry S. Upshaw, Leah Hart, Bill Michael, Mike Strube, F. DeWitt Kay Jr., Jack Yurkiewicz, Gary Salton, Frank Montague, Jim DeArmon, Stephen Cole, Scott Millis, Bob Weiss, Sharon Foster, Ton Stam, David Chen, Charles Bennett, Ahsan Arozulla, Jim Adams, and Sul Kassicieh. Of course, our friends at the American Psychological Association Books Department—including Ed Meidenbauer, Peggy Schlegel, Julia Frank-McNeil, Mary Lynn Skutley, Dan Brachtesende, and Gary R. VandenBos—took the risk and provided our book contract. We extend special thanks to Alan Phillips for providing Programmer's File Editor for this package. We would be remiss not to acknowledge the resources, including consulting and access to computers that were provided to us in our pursuit of the ODA paradigm by several outstanding academic computer centers. In this context we wish to thank the National Supercomputer Facilities at Champaign and Cornell, the Advanced Computing Research Facility at Argonne National Laboratory, and the Computer Center at the University of Illinois at Chicago. We also wish to acknowledge financial support for our early research provided by Ron Manaster, Goodman Manaster, Inc. Finally, our decision to create the ODA systems served to disrupt the lives of our friends and colleagues. We wish to thank all of the members of our social networks for their understanding and support in this context. However, as is true in most any birthing process, it was our families and we who were most incessantly and directly affected by this project. We wish to express our deepest appreciation to our loved ones for their patient, optimistic, and loving support.

Optimal Data Analysis

CHAPTER 1
Introduction to the ODA Paradigm

ODA—pronounced with a long "O" sound ("oh-dah")—is the short way of referring to the "optimal data analysis" paradigm. This new statistical paradigm is simple to learn, the software is easy to operate, and the findings of ODA analyses have intuitive interpretations. Nonetheless, the paradigm and the software are powerful and rich. This book describes the ODA paradigm and software and demonstrates how to apply ODA in the analysis of data. Everything needed to understand ODA is contained within this book, and all ODA analyses within this book can be accomplished using the accompanying software. Only moments away from jumping directly into the fray, we pause briefly to address four questions frequently asked by beginner and expert alike as they ponder the merits of learning ODA.

What Is ODA?

ODA is a new statistical paradigm—a quantitative scientific revolution, so to speak. Perhaps the best way to illustrate what this means is by example.

The ordinary least squares (OLS) paradigm maximizes a *variance ratio* for a given sample, and includes analyses such as t test, correlation, multiple regression analysis, and multivariate analysis of variance. If one wishes to maximize a variance ratio, then the OLS paradigm is required, obviously, because that is what it does. That is, maximizing variance ratios is what the "formulas" that compute t, F, and r actually accomplish for a given sample (Grimm & Yarnold, 1995, 2000).

In contrast, the maximum likelihood (ML) paradigm maximizes the *value of the likelihood function* for a given sample. This paradigm includes analyses such as chi-square, logistic regression analysis, log-linear analysis, and structural equation modeling. If one wishes to maximize the value of the likelihood function, then the ML paradigm is required (Grimm & Yarnold, 1995, 2000).

In contrast, ODA maximizes the *accuracy* of a model. As a simple example, imagine we wished to assess whether two groups—Group A and Group B—of independent observations can be discriminated on the basis of their score on a test. ODA identifies the model that uses the test score in a manner such that it discriminates members of A versus B with theoretical maximum possible accuracy. To understand how this is accomplished, recall that the model—*any* model, *every* model—can actually physically be used to compute each observation's "score" via an equation or "formula." The resulting score is then considered with respect to the decision criteria of the particular procedure and a prediction is then made on the basis of the model. In the present example, in some instances the model (regardless of the methodology by which it was developed) will predict that an observation is from Group A. Other observations will be predicted to be from Group B. Every time the predicted group membership status of an observation is correct—the same as the actual group membership status, a point is scored. An incorrect prediction scores no points. Obviously, the largest number of points that it is possible to attain for a sample of N observations, in theory, is equal to N, the number of observations in Groups A and B that are classified by the model. Clearly, this maximum score is only possible if all observations are correctly predicted to be from A or B by the model. The minimum score possible is obviously zero points, in which case all observations are *incorrectly* predicted to be from A or B by the model.

By definition, an ODA model achieves maximum possible accuracy for a given sample of data, in the sense that no other model that is based on the test score can achieve a superior number of points. All possible alternative models are (explicitly or implicitly) evaluated to literally prove this, which is one reason why ODA is "computationally intensive." As OLS maximizes a variance ratio for a given sample of data, and as ML maximizes the value of the likelihood function for a given sample of data, ODA maximizes the accuracy of the model for a given sample of data. Of course, if different observations can be weighted by a different number of points, for example if a "natural" weighting metric, such as time, weight, or cost is available, then weighted accuracy may be maximized (or cost minimized), as may be desired by the operator.

Every type of analysis (i.e., every specific configuration of data, constraints, and hypotheses) that can be conducted in the OLS and ML paradigms can be conducted in the ODA paradigm. The ODA paradigm can conduct many analyses that one simply cannot do using either OLS or ML paradigms. ODA is much more general, and much more encompassing of different data, constraints, and hypothesis configurations, than are the alternative statistical paradigms. The ODA paradigm is, quite literally, "new and improved." Using this paradigm, and only using this paradigm, is one able to identify maximally accurate models for a given sample.

Why Is ODA Superior to Other Data Analysis Programs?

The ODA paradigm is vastly superior to alternative paradigms. Consider first, *conceptual clarity*. For every problem analyzed via ODA there is one precise, optimal analysis—a specific given data configuration and hypothesis dictates the exact nature of the ODA model that is appropriate. Using traditional statistics, in most applications several different analyses are feasible—all reflecting some degree of lack of fit between their required underlying distributional assumptions and the actual character of the data. Consider second, *ease of interpretation*. Every ODA analysis provides the same intuitive goodness-of-fit index: for every ODA analysis an index is computed on which 0 reflects the accuracy expected by chance for the sample, and 100 reflects perfect accuracy. Using traditional statistics, different analyses provide different goodness-of-fit indices that are non-intuitive and that are not directly comparable across procedures. Consider third, *maximum accuracy*. Every ODA analysis provides a model that guarantees maximum possible accuracy. Using traditional statistics, no analysis provides a model that explicitly guarantees maximum possible accuracy. Consider fourth, *valid Type I error*. No ODA analysis requires any simplifying assumptions, and p is always valid and accurate—a permutation probability derived via Fisher's randomization method, invariant over any monotonic (i.e., transformed values either always increase or always decrease) transformation of the data. Traditional analyses require simplifying assumptions (e.g., normality), p is only valid if the required assumptions are true for one's data, and p may be inconsistent over transformations of the data.

An obvious advantage of ODA software is *availability* (the first and currently the only software available that performs ODA). There are many good packages available for performing OLS and/or ML analysis, and many are an order of magnitude more expensive than the ODA book/software. Comparing software across paradigms, the ODA software is superior to software of earlier paradigms for two important reasons. Consider first, *ease of learning and teaching*. Everything needed to understand the ODA paradigm and analyze data is discussed in this book. Many courses, books, and articles are needed to understand traditional statistics and to correctly operate associated software, requiring years of study. Consider second, *ease of use*. Most types of ODA analyses require the same basic set of seven programming commands. Using traditional procedures requires learning of numerous—hundreds—system-unique programming commands.

Who Is the Audience for This Book and Software?

A "must-have" tool for all quantitative researchers, ODA is the only general-purpose statistics software that explicitly maximizes predictive accuracy. Many articles were published during the software's beta testing in areas such as developmental psychology, pediatrics, social psychology, physiology and physiological psychology, allergy-immunology, cardiology, emergency medicine, clinical psychology, criminal justice, education, industrial–organizational psychology, political science, social work, sociology, economics, AIDS research, women's studies, biology, general internal medicine, psychiatry, management science, rehabilitation medicine, neurology, pharmacy, marketing, and oceanography. Conceptually straightforward examples in the book represent a myriad of substantive areas drawn from many disciplines, and many examples can only be solved using ODA. Regardless of discipline, researchers ranging from new students to seasoned professionals will marvel at how easily and rapidly ODA theory and software can be mastered—and can help them master their data. Knowledge needed to conduct and interpret a cornucopia of different statistical analyses is provided in easy-to-follow steps in the book—the "no formula" presentation is engineered to maximize conceptual clarity. Similarly, the "consistent-across-design" software encourages ease of learning as well as efficiency—because most analyses require the same basic set of seven commands. The combination of ultrarapid learning and maximum accuracy results, made possible only by ODA, promises levels of efficiency that researchers only dream of. Welcome to the new revolution in statistics!

It is recommended that the person wishing to master this book and software system have familiarity with some basic statistical concepts (e.g., what is a variable or what is a sample of independent observations), and some basic computer skills (e.g., how to copy files from a CD to a new directory). No prior knowledge of any statistical procedures is required to master this book and software. On the other hand, the deeper one's understanding of alternative procedures and software systems, the greater one's appreciation for the ODA paradigm and software. For all users, the more one uses the software, and the more different types of data that one analyzes, the more one will appreciate the conceptual clarity and ease of use of ODA.

How Should the Reader Use This Book?

In our opinion, the best way to learn and master the ODA paradigm is to read this book systematically (begin at the beginning), work each sample problem when it is presented, and then pause before continuing in order to use your own (real or artificial) data to practice each particular method. If you do not have data for the example at hand, another option is to manipulate the data that are provided for the example and note the effect of your manipulation on the results. This book is linear—every page builds on ideas presented in prior pages. At every point at which an analysis technique is discussed, each technique may, of course, be used to analyze unpublished scientific data, and the findings published in leading scientific journals in any substantive area (we do plenty of this). Of course, data already published using other methods may be reanalyzed using ODA, and comparisons of the findings published in leading scientific journals, in substantive applied journals, or in applied statistics and methodology journals (we do plenty of this). Some people will apply this software to data in ways that are not yet covered in this book, but that *will* be covered in a future edition of this book once we read their papers.

At this point we are ready to begin. We hope that you have decided to take this journey, as we have little doubt that you will be pleased when you complete it. We begin at the beginning.

Basic Steps and Key Concepts

The first step of any ODA analysis is to define what it is that you wish to predict. In ODA, a *class variable* is any random variable that may attain two or more levels: the levels reflect the phenomena that one desires to predict. In conventional statistics, class variables are often referred to as dependent variables. Example class variables include health status (sick, healthy), socioeconomic status (lower, middle, upper), or change in price of a commodity futures contract (lower, unchanged, higher). The *category level* of a class variable is the number of different values or levels that the class variable may attain. Thus, as defined above, health status is a two-category class variable, and socioeconomic status and investment outcome are three-category class variables. During analysis, class categories are identified via *dummy-codes*: Class 0, Class 1, Class 2, and so forth.

In the best of all worlds, the class categories should represent qualitatively distinct phenomena, conditions, or states. For example, biological sex—male versus female—involves two qualitatively different categories, and is a highly stable class variable. Other variables, however, are less ideally suitable. For example, imagine that one wished to predict mortality status: alive versus dead. Further imagine that, to do this, a sample of patients was prospectively followed for one year, after which time patients were classified as either dead or alive. It is possible that, had the study been continued for one additional day, some of the patients classified as being alive would instead have been classified as dead. Another example of an imperfect class variable is age category. For example, suppose that we are interested in comparing geriatric (65 years or older) versus non-geriatric (younger than 65 years) people.

Further imagine that a participant in our study will turn 65 years of age in one more day—or in two more days. The greater the potential instability or unreliability of a class variable, the more "fuzzy" it is. Instability in the class variable that occurs near the cutpoint (e.g., 65 years in this example) is problematic, theoretically limiting the upper bound of accuracy that it is possible for an ODA model to attain.

Finally, pragmatically speaking, it is a good idea for those learning the ODA paradigm to begin by studying class variables having two category levels: so-called dichotomous or binary class variables. This is because ODA is computationally intensive—problems become much more difficult to solve as the number of category levels increases. Furthermore, in the event that one's ODA model is unable to perfectly predict or nearly perfectly predict a binary class variable—which unfortunately is usually true—one does not want to increase the complexity of the problem by including additional "shades of gray." Of course, it is possible that adding an intermediate category—"undecided"—might enhance model performance in those instances in which some participants cannot be reliably classified into either type of the dichotomy (Dr. Loretta J. Stalans, personal communication, 2002).

The second step of any ODA analysis involves defining the set of potential predictor variables. In ODA an *attribute* is any random variable that can attain two or more levels. Attributes are used to predict the class variable. In conventional statistics, attributes are often referred to as independent variables. In addition to defining the attributes, one must identify their *metric* (cf. Velleman & Wilkinson, 1993). For ODA the primary distinction is *qualitative* versus *ordered* attributes. Qualitative attributes may be *binary*—such as smoking status (smoker, non-smoker), or *polychotomous*—involving three or more qualitatively distinct, unordered categories: occupation (unemployed, clerk, lawyer) or investment decision (buy, hold, sell), for example. Ordered attributes—for which increasing scores indicate increasing values of the phenomenon, may be *ordinal* (e.g., a rating made using a 7-point Likert-type scale), *interval* (e.g., score on a college board examination), or *ratio* (e.g., time). Finally, it is important to determine whether there is an *a priori hypothesis* relating the attribute and the class variable (discussed ahead). Of course, as is true in conventional statistics, the decision concerning which variable serves as an attribute and which serves as a class variable is usually arbitrary. For example, in studying substance abuse in schizophrenia, Mueser et al. (1990) treated substance abuse as the independent variable, and Mueser, Yarnold, and Bellak (1992) treated substance abuse as the dependent variable.

The third step of any ODA analysis involves specification of appropriate *weights:* weights are used so that an obtained ODA model mirrors reality. For example, if we wished to obtain a model to guide stock market investment decision-making, then we should weight the observations (i.e., the different days in the study period) by the amount of money that the stock went up or down—because we wish the model to be most accurate on days that the price changes substantially—in order to maximize profit. Without weights, the ODA model would simply maximize the number of correct decisions: the nonweighted model might get more buy/sell decisions correct than the weighted model, but the weighted model could still be much more profitable by emphasizing accuracy on highly volatile days.

There are two types of weights. The first type of weight is called *prior odds*: analogous to the use of *antecedent probability* or *base rate* in Fisher's discriminant analysis, weight all n_c observations in class category c by the value $1 / n_c$ (e.g., Greenblatt, Mozdzierz, Murphy, & Trimakas, 1992; McLachlan, 1992; Meehl & Rosen, 1955; Rorer & Dawes, 1982; Widiger, 1983). For example, imagine that we wished to predict whether a patient hospitalized with pneumonia would survive. If we obtained an ODA model for predicting mortality attributed to pneumonia, but failed to consider the base rate for mortality among hospitalized pneumonia patients, the model might overestimate the number of patients who died (most cases of common pneumonias are not fatal). Later it will be shown how one can estimate the classification performance obtained by one's final model for all possible base rates (this is known as assessing the *efficiency* of one's model). Then, to determine how well the model will perform in a given sample (e.g., zip-code-defined geographic area), one need only consider the base rate for that sample and consult the curve derived in the efficiency analysis.

The second type of weight is a quantitative assessment of the value or importance of the attribute to the decision-maker. For example, in an application involving predicting daily movement in the price of a stock, we would weight observations by the dollar value of the change in stock price. Were we to construct an ODA model for this application without considering the return, the model would maximize our ability to predict the direction of movement of the stock correctly. In the absence of a return weight, we might be correct, for example, 85% of the time that we predict a movement in stock price—and yet still lose money because the model misclassified the days on which the price of the stock changed the most. However, specification of a return weight would obtain an ODA model that maximized the amount of dollars correctly predicted: although overall predictive accuracy might decrease (e.g., to 40% correct predictions), the model would seek correct prediction of the times that the stock value changed substantially, and thereby maximize profit.

Functionally, ODA models are maps between values on the attribute and predicted class memberships. An example of an ODA model for a qualitative attribute is: If the person is a smoker, then predict disease; if the person is a non-smoker or an ex-smoker, then predict health. An example of an ODA model for an ordered attribute is: If the person smokes 4 or more cigarettes per day then predict disease; otherwise predict health. Essentially, an ODA model is a decision rule for predicting class membership status on the basis of the attribute. For a given (*training*) sample of data, this rule yields the theoretically maximum-attainable level of (weighted or nonweighted) *percentage accuracy in classification* (abbreviated as PAC). An observation is *correctly classified* when the actual and predicted class membership are the same. For example, the model predicts death and the person dies. An observation is *misclassified* when the actual and predicted class memberships differ. For example, the model predicts death and the person lives. The number of correctly classified observations in the sample is called the *optimal value*. The theoretical maximum possible value for both weighted and nonweighted PAC is 100%.

Historical Perspective

An application involving a single attribute is referred to as a univariable ODA (UniODA) problem, and an application involving more than one attribute is known as a multivariable ODA (MultiODA) problem. Initial exploration of UniODA occurred in the 1950s, as researchers began to address the problem of how to best assign observations into one of two mutually exclusive categories on the basis of their score on a single test or on a single composite index based on several tests. As is true of modern UniODA methodology, early research began by considering that for a two-category, single attribute problem, observations are assigned to one or the other category on the basis of a cutting score (CS). Observations with scores on the test that exceed the CS are assigned to one category, and observations with scores that fail to exceed the CS are assigned to the other category. An optimal cutting score (OCS) has the property that, compared to other possible CSs, the OCS results in the maximum proportion of correct decisions.

The earliest programmatic discussions of the definition and use (for classification purposes) of the OCS that we have located were presented by Duncan, Ohlin, Reiss, and Stanton (1953) and by Meehl and Rosen (1955). They addressed modern UniODA concepts, such as maximization of overall classification accuracy; maximization of the sensitivity of the model for a user-specified category of the class variable; the need for weighting by prior odds in imbalanced applications involving different numbers of observations in the different categories of the class variable; and constrained optimization (see also Alf & Abrahams, 1967; Blumberg, 1957; Dawes & Meehl, 1966; Luce & Raiffa, 1957; Rosen, 1954). However, the computational effort required to obtain an OCS by hand rendered this procedure infeasible for most real-world samples, so a suboptimal normal-based heuristic for obtaining an OCS was developed (Cureton, 1957; Darlington & Stauffer, 1966a) that allowed weighting either by prior odds (Darlington & Stauffer, 1966b; Dawes, 1962; Golden & Meehl, 1979; Rorer & Dawes, 1982; Rorer, Hoffman, LaForge, & Hsieh, 1966) or misclassification cost (Rimm, 1963; Rorer, Hoffman, & Hsieh, 1966).

Research in these areas declined steeply in the 1970s and remained dormant until the late 1980s. Although the reason(s) underlying this decline are not fully understood, we speculate that two important contributing factors included (a) the difficulty involved in hand-computing the OCS, particularly in weighted applications; and (b) a shift in focus—as computers became more accessible and capable—to the dramatically different and more difficult problem of obtaining MultiODA solutions (see Epilogue). Because of the enormous computational resources required to solve MultiODA problems, most of the latter research involved relatively limited experimental comparisons of the classification performance achieved using a MultiODA model versus suboptimal multiattribute classification models such as might be obtained using logistic regression or Fisher's discriminant analysis. It is thus not improbable that the failure of prior research to discover the broader implications of the OCS concept—that is, as concerns the existence of the ODA paradigm—was due in large part to the unavailability of an efficient methodology for solving ODA problems.

The recent discovery that exact statistical distributions underlie ODA models—which are so flexible that the models may be specified to precisely reflect the finest details in any experimental design—paved the way for the discovery of ODA as a general statistical paradigm. However, the basic methodology used to identify ODA models—that is, models that specifically maximize return or minimize cost for a given scenario—is used in a variety of academic disciplines. For example, the basic technical concept involved in identifying the optimal (most accurate) ODA model is known as maximum feasible subsystems of linear inequalities (Max FS), along with the closely related concept of minimum irreducible infeasible subsystems (Min IIS). Both Max FS and Min IIS have been appearing with increasing frequency in the literature of operations research (Chinneck, 2001; Chinneck & Dravnieks, 1991; Gleeson & Ryan, 1990; Mangasarian, 1994; Marcotte & Savard, 1995; Soltysik & Yarnold, 1994b; Van Loon, 1981), computer science (Amaldi, 1995; Amaldi, Pfetsch, & Trotter, 2003), machine learning (Bennett & Bredensteiner, 1997; Parker & Ryan, 1996), and perceptrons and neural networks (Hastad, 2001; Mattavelli & Amaldi, 1995). Examples of recent applications include radiation therapy planning (Sadegh, 1999), speech translation (Kussner & Tidhar, 2000), computational biology (Wagner, Meller, & Elber, 2002), and digital television broadcasting (Rossi, Sassano, & Smriglio, 2001). Clearly, the stage is now set for the application of this new paradigm in statistical analysis to a host of empirical data.

Thirty Hypothetical Applications

What types of problems can ODA solve? To begin to answer this question and illustrate the flexibility of the approach, consider the following hypothetical applications for which ODA is the optimal analytic methodology.

Astrology

Imagine that one is interested in optimally predicting personality differences between people born under different astrological signs. To address this issue, one might conduct a study in which a large random sample of persons whose zodiac signs were known all take a series of personality tests. With these data ODA can determine, with maximum possible PAC, (a) which signs of the zodiac are associated with relatively high or low scores on the different personality tests relative to other signs of the zodiac, and (b) exactly what values on the personality tests to use as an operational definition of relatively high versus low scores. Other attributes, such as dominance or nurturance, have also been hypothesized to differentiate people born under different astrological signs and could be investigated. ODA can also determine whether the findings of this research are consistent for different data samples, such as reflected by data collected from different races, genders, religious orientations, and socioeconomic status levels. If the results are not consistent across all samples, ODA can determine which samples (if any) share which ODA model.

Astronomy

When will there be relatively heavy meteorite showers? To address this question, one might conduct a study in which data are collected from Hawaiian observatories for three years. Each day, the presence or absence of unusually active meteorite showers is recorded. For the sake of illustration, imagine that the criterion for unusually heavy showers is more than 15,000 meteorites per hour in one's observation area. In addition, the level of solar flare activity is recorded (a time lapse might be appropriate) as the attribute. With these data ODA can determine what level of solar flare activity predicts heavy meteorite showers with maximum PAC. If the actual number of meteorites per hour is recorded, ODA can determine what level of solar flare activity optimally predicts the number of meteorites per hour. If the mean mass of the meteorites can be validly estimated, ODA can determine the level of solar flare activity that optimally predicts meteorite mass per hour. Other attributes, such as the level of sunspot activity, Earth's magnetic activity, or the distance of the Earth from the sun, moon, other planets, or asteroid belts, might be investigated. ODA can also determine whether findings generalize to other observatories, and can identify groups of observatories that share a consistent ODA model.

Beer Brewing

What is the best recipe for beer? To address this issue, one might conduct a study in which data are collected from a random sample of beer drinkers. Each observation is randomly assigned to try one of three different brews of beer that differ only in the weight of hops added to the recipe: one pound, two pounds, or three pounds of hops. Observations rate whether the beer tastes good or bad. With these data ODA can determine, with maximum possible PAC, which brew(s) are associated with good versus bad ratings, and which brew(s) maximize the percentage of the sample responding with good evaluations. If observations also provide a numerical rating, say on a 10-point scale, of the desirability or goodness-of-taste of the beer, ODA can determine which brew(s) are associated with maximum overall satisfaction. If the amount of beer consumed by the observations is recorded, ODA can determine which brew(s) are associated with maximum consumption. Other attributes that might predict taste ratings, such as the type of water used, method of fermentation, length of storage, effect of different combinations of other ingredients in the recipe, and alcohol content of the beer could also be investigated. ODA can also determine whether the findings generalize to other data samples, such as different races, genders, and ages. If the findings are inconsistent, ODA can determine which samples share consistent ODA models.

Bird Watching

Imagine that one is interested in determining the optimal time(s) in the morning for spotting endangered birds. To address this issue, one might conduct a study in which data are collected on the hour between 5 a.m. and noon every day for two months. Each hour a survey of a wildlife refuge is conducted. If at least one endangered bird was seen, the hour is deemed a success; if no endangered birds were seen, the hour is deemed a failure. With these data, ODA can determine the optimal time(s) in the morning to spot endangered birds. If the total number of endangered birds seen each hour is recorded, ODA can determine the time(s) in the morning in which the maximum total number of endangered birds can be seen. If the total number of different endangered species seen each hour is recorded, ODA can determine the time(s) in the morning in which the maximum variety of endangered birds can be seen. Other attributes that might be investigated include the air temperature, wind velocity and direction, presence of precipitation, presence of other animals, or the amount of ground cover. ODA can also determine whether findings are consistent for different genders or species of bird for different seasons. If the findings are inconsistent across different samples, ODA can determine which samples (if any) share which ODA model.

Credit Collection

Imagine that one is interested in accurately predicting which of a pool of observations might best be hounded for past-due bill payment. To address this issue, one might conduct a study in which data are collected for a comprehensive sample of all customers of a large department store over a span of eight consecutive months. The class variable is whether the observation paid the outstanding bill by the end of the study period. Imagine that the observation's age is the attribute. With these data, ODA can determine the age(s) that best predict bill payment. If the amount of the bill is recorded, ODA can determine the age(s) that pay the greatest amount or percent of money. Other attributes, such as type of solicitation message, marital status, income, or past credit history might also be analyzed. Combinations of these and other attributes may be used to define new class variables, or be treated as multiple samples to evaluate consistency.

Credit Screening

Imagine that one is interested in minimizing the need for credit collection, and desires an optimal model for determining the credit-worthiness of applicants for credit cards or loans. To address this issue, one might conduct a study in which data are collected for a comprehensive consecutive sample of 200 applicants for a credit card, all of whom were accepted for the purposes of the study and then followed for one year. At the end of the year it is determined whether each observation had been a positive or negative revenue source for the lender (class variable). Following the lead of the preceding example, the attribute is the age of the observation. With these data ODA could determine the age(s) that attain positive or negative status with greatest relative frequency. If the amount of the profit, or the absolute value of the loss, associated with each observation is recorded, ODA can determine the age(s) that return the greatest amount (or percentage) of money. Other attributes, such as gender or ethnicity, marital or parental status, religious or political affiliation, or income or past credit history might also be analyzed—so long as it is legal and ethical. Combinations of these and other attributes could also be used to define new class variables, the generalizability of the findings could be evaluated, and consistent samples could be identified.

Criminal Justice

Which attributes accurately predict whether a defendant is convicted in a criminal trial? To address this issue one might conduct a study in which data are collected from random samples of 50 criminal trials from each of 10 counties in a single state. The attributes might include gender and age of the prosecutor, defender, and client, type of evidence emphasized, presentational manner (e.g., cold and calculating versus animated and dramatic), prior record of the defendant, characteristics of the judge or jury, and type of crime. With these data, ODA can determine the ability of each attribute to accurately predict the disposition of the trial. At the user's discretion, models may be determined that obtain optimum PAC over all 10 counties when they are considered as a single sample, or, alternatively, that obtain the optimum mean PAC when applied to each of the 10 counties individually. Were the time taken to decide on the guilt or innocence of the defendant recorded, or to conduct the trial, ODA could find the model that best predicted relatively fast or slow decisions. Were cost data available, ODA could find the model that best predicted expensive, inexpensive, or cost-effective (defined by the user) trials. Were sentencing information available, ODA could find the model that best predicted the severity (length) of sentences. Were recidivism (returning to a life of crime after being released from prison) data available, ODA could find the model that minimized the relative frequency of recidivism, or the weighted cost of the recidivism (e.g., in dollars for crimes committed against property; in lives or rapes for crimes committed against people). And, were data available concerning reintegration of the survivors of this study into mainstream society, ODA could help identify variables that best predict success and that best predict failure in this context.

Dating

Imagine that one is interested in discovering factors that best predict one's personal dating bliss. To accomplish this, one might conduct a study in which a person collected data concerning all dates (observations) occurring over some designated time period. The class variable is an indication of whether one desired to redate the observation. A host of personally salient variables might serve as attributes. ODA can determine the ability of each attribute to optimally predict the desirability of one's date. Also, dates may be rated by any desired subjective or objective measure, and ODA used to discover the attributes that best predict the weighted satisfaction. Note that this methodology can be used to determine an optimal model for any desired personally salient issue. Also, investigators, counselors, educators, and other professionals who work with single cases could use this method to discover statistically reliable markers of change, for example, in scale scores or behavioral observations of behavior, in single-case longitudinal series (cf. Yarnold, 1992).

Direct Mail Advertising

How can one determine the most productive people to whom to direct-mail an advertisement? To address this issue, one might conduct a study in which data are collected from a random sample of people about whom psychoethnographic data (attributes) were available, and who were mailed an advertisement. Six months after the mailing, observations would be coded as either having responded or not responded to the advertisement. ODA could be used to optimally evaluate the ability of each attribute to predict whether observations responded to the advertisement. If the amount of money each observation spent is recorded, ODA could be used to optimally evaluate the ability of each attribute to predict the return of the mailing. Data from different samples, such as reflected by different product lines, could be used to evaluate the generalizability of the findings, and samples for which consistent findings emerged could be identified.

Driver Licensing

Imagine that one is interested in improving the quality of licensed automobile drivers (to the extent possible) by optimally determining the value constituting the minimum passing score (MPS) on the written examination section of the licensing application. To address this issue, one might conduct a study in which data are collected from all people who lived and worked in a given county, and who applied for and received a driver's license during the past year. Each observation would be tracked using mailed questionnaires, telephone interviews, and police and insurance files for one year after receiving the license. It would be recorded whether, during the year, a parking violation or moving violation citation, accident, injury or fatality report had been generated naming the observation as the culpable party. Using these class variables, ODA could determine the value of the MPS that resulted in optimal prediction of any of the class variables, or in optimal prediction of any of the categories constituting any of the class variables (e.g., specifically, fatalities). Had the number of each event reflected by the class variables, and/or their corresponding direct (repairs, litigation) or associated (time off from work, physical and emotional trauma) costs been recorded, ODA could determine the value of the MPS that resulted in optimal prediction of the number or cost of any of the class variables/categories. Other attributes that might predict the class variables, such as age, geographic region (urban, rural), or prior driving record might also be investigated. Were data available from multiple samples, such as reflected by data collected from multiple counties or from professional drivers, ODA could optimize the models for the pooled data or simultaneously and separately across the samples and could identify the samples for which the findings were consistent.

Epidemiology of AIDS

How can one discover factors that predict progression from testing positively for HIV infection to development of AIDS? To address this issue, one might conduct a study in which data are collected from a comprehensive consecutive series of individuals who tested positively for HIV infection at a clinic. People are tracked for exactly one year following initial diagnosis, at which time it is recorded whether each person had AIDS (the class variable). The attribute of greatest interest is the number of T4 initiator cells. ODA can determine the critical number of T4 initiators that optimally predicted whether a person in this sample developed AIDS within the first year of diagnosis of HIV infection. A host of other attributes, such as drug use, blood chemistry measures, dietary, physical, and sexual behaviors, psychoethnographic measures, and so forth could also be used to predict the development of AIDS. Were data concerning hospitalizations, fatalities, or treatment costs for each person recorded, ODA could determine the number of T4 initiators that optimally predicted the number of hospitalizations or fatalities or the cost of medical care. Were data collected from multiple clinics, the generalizability of the results could be determined, and samples with consistent findings could be identified. Were data from many clinics available, ODA could be used to discover factors that optimally discriminate clinics with consistent findings in the prior analysis from clinics with inconsistent findings.

Farming

What is the optimal amount of cow manure to mix into soil to ensure that as many watermelons as possible grow to at least twenty pounds? To address this question, one might conduct a study in which the land available to conduct this research is randomly subdivided into 25 equivalently sized plots, each sufficiently large to support ten mature vines. Five different levels of manure (5%, 10%, 15%, 20%, and 25%) are to be contrasted, and five plots of land are randomly assigned to (and mixed at) each manure level. At the end of the growing season, all watermelons are harvested and weighed. If a watermelon achieves a weight of twenty pounds or more, it is classified as a success; otherwise, it is classified as a failure. ODA can determine the level(s) of manure that optimally predict success versus failure or that predict either successes or failures with maximum PAC. ODA can also determine the level(s) of manure that optimally predict the total weight of watermelons harvested. If the sales prices of the watermelons are recorded, ODA can determine the level(s) of manure that optimally predict the total return on the harvested watermelons. Other attributes, such as the amount of water or direct sunlight or the brand of seed used could also be investigated. Data from other samples (e.g., other cash crops) could be used to evaluate the generalizability of the findings, and ODA could identify samples with consistent ODA models.

Fishing

What is the optimal lure color for catching fish? To address this issue, one might conduct a study in which data are collected for 100 hours of fishing. Each hour, a gold, silver, red, yellow, or green lure is randomly selected and used: If any fish were caught during this hour, the hour is deemed a success; if no fish were caught during the hour, then it is deemed a failure. With these data, ODA can determine which lure(s) to use in order to maximize fishing success. If the number of fish caught each hour is also recorded, ODA can determine which lure(s) to use in order to maximize the number of fish caught per hour. If the weight of each fish is also recorded, ODA can determine which lure(s) to use in order to maximize the total weight of the fish caught per hour. In addition to lure color, other attributes that might be investigated include the type of lure, time of day, air or water temperature, barometric pressure, depth of the water, weight and type of fishing line used, speed of the retrieve of the lure, or any other similar attribute. ODA can also determine whether the findings are consistent for different data samples, such as reflected by data collected from different lakes, different seasons, or different species, ages, and gender of fish. Finally, if the results are inconsistent across samples, ODA can determine which samples (if any) share which ODA model. Note that this example also works well with hunting of both animals and edible wild foods.

Gambling

Imagine that one is interested in evaluating the validity of several types of information as predictors of whether a horse will finish in one of the top three places ("in the money") in a race. To address this issue one might conduct a study in which data are recorded for all races occurring over a one-year period at a nearby track. For each race, horses are classified as either finishing in the money or not (class variable). Attributes include the number of times the horse finished in the money in the last two months, the win/loss record of the jockey in the past week, and the workout speed of the horse on the day of the race. ODA could be used to optimally determine the ability of these attributes to predict whether a horse made money. If the actual amount of money the horse won was recorded, ODA could optimally determine the ability of these attributes to predict the amount of money won by a horse. ODA could also be used to assess the generalizability of these findings across different racetracks. Conceptually similar methods could be used to determine which money market or mutual fund is the best investment.

Golfing

Imagine that one is interested in improving one's putting. To address this issue one might conduct a study in which data are collected for 500 consecutive putts made at the local nine-hole golf course. The class variable is whether the attempted putt was successful, and the attribute is the type of putter used (standard versus long-handled). For each putt, the putter used is randomly selected. ODA could be used to determine the optimal choice of putter in order to maximize one's putting success. If the length of each putt was recorded, ODA could determine the optimal choice of putter for maximizing the number of long- or short-range successful putts made. Other attributes, such as the type of grip, stance, or aiming protocol could also be studied. The findings could be generalized across different samples such as different courses or types (e.g., flat or curvy) of greens, and samples for which consistent findings emerged could be identified.

History

Imagine that one is interested in determining whether the relative frequency of reports concerning different types of aberrant interpersonal relationships published in the local newspaper fifty years ago is consistent with reports published in that newspaper today. To address this issue one might conduct a study in which the local paper is obtained for the entire years of both 1942 and 1992. Every article concerning negative (by Western standards) interpersonal events (divorce, adultery, beatings, robberies, murders, rapes) is pulled. With these data, ODA could determine whether the relative frequencies of such events within time were consistent across time, thus reflecting historical homogeneity. ODA might also discover events for which the 1942 and 1992 relative frequencies are inconsistent, reflecting historical heterogeneity. Were data collected from multiple samples, such as different newspapers or books, ODA could determine the extent to which the findings generalized, and could identify samples for which consistent findings emerged. It should be noted that one can use this technique to discover distinguishing features, if they exist, between (time-lapsed or contemporary) competing entities on the basis of the comparability of the corresponding relative frequency of the different aspects (features) that define their constitution. ODA can then be used to optimally contrast the relative frequency of the content (aspects or features) offered by, for example, different television shows, colleges, resorts, or mountain ranges.

Hostage Negotiation

Imagine that one is interested in evaluating the relative efficacy of different strategies for negotiating for the safe return of hostages. To address this issue, one might conduct a study in which data are collected for all hostage situations occurring in the United States during 1991. The class variable is whether any hostages were killed, and the attribute is whether the authorities used force in an attempt to free the hostages. With these data, ODA could determine the optimal ability of the use of force to predict whether any hostage lives would be lost. Had the number of hostage deaths been recorded, ODA could predict with maximum accuracy the number of hostage deaths resulting from the use of force. Other class variables (e.g., presence and/or amount of property damage or of kidnapper casualties; whether the situation involved kidnapper relatives) and attributes (e.g., political affiliation, number, gender, age, and cross-cultural communication and knowledge status of the kidnappers; type of demands; type and/or quantity of weapons; nature of the area where hostages are being held) could also be investigated. ODA could also assess the generalizability of the findings across multiple samples (e.g., different countries or different time periods), and, if findings were not consistent over all samples, ODA could identify samples for which consistent findings emerged.

Hurricane Forecasting

Imagine that one is interested in determining factors that predict with maximum PAC whether a hurricane will falter at sea or come crashing ashore. To address this issue one might conduct a study in which a sample of hurricanes is tracked from their inception (note that the development of full-fledged hurricanes from tropical storms constitutes an interesting class variable). Assessed at both the eye and periphery, attributes include hourly measurements of the storm speed, height, surface area, and cubic volume; direction and sustained velocity of the storm winds and of the steering winds and currents; strength, direction, and temperature of opposing frontal weather systems; water temperature and depth (and change in water temperature and depth); direction and strength of tide and saline thermoclines; location and phase of the moon; and time of day. ODA can determine the optimal ability of each attribute to predict whether a storm will come ashore. If the category of the storm (Category I [least severe] to Category V [most severe]) were recorded, ODA could optimally predict the strength of the storms that came ashore. Were data concerning the cost of the property and other resources destroyed by the hurricanes recorded, ODA could optimally predict the cost of the storms that came ashore. Of course, ODA could evaluate the generalizability of the findings across, for example, different oceans or hemispheres. Conceptually related research might also focus on other natural disasters, such as avalanches, blizzards, droughts, earthquakes, fires, floods, tornadoes, tsunamis, or volcanic eruptions. Although less exotic, such methods may facilitate optimal forecasting of precipitation (rain, sleet, snow), or of changes in weather phenomena, such as temperature, pressure, winds, or humidity. Finally, similar methods might be fruitfully used in forecasting the emergence (etiology) and spread (ontogenesis) of biological disasters, be they from land (army ant, locust), air (killer bee, mosquito), or sea (lamprey, zebra mussel).

Life Insurance

Imagine that one is interested in determining the most desirable clients (observations) to whom to attempt to sell life insurance. To address this issue, one might conduct a study in which data are collected from all people who were interviewed—in the course of regular business—by a specific insurance agent during the past six months. The class variable is whether the observation bought life insurance. Attributes that might predict the class variable included, for example, the observations' gender, parental status, age, socioeconomic status, education, occupation, perceived health status, recent experience with death, or state, trait, or somatic anxiety. Using ODA, one may determine the optimal ability of each attribute to predict whether an observation will purchase insurance. With data concerning an observation's corresponding net return (i.e., payments − costs), ODA could determine each attribute's optimal ability to predict the overall return of the insurance sold. A conceptually related problem would involve maximizing the prediction (or minimizing the occurrence) of delinquent accounts or catastrophic diseases/accidents.

Missionary Work

Imagine that one is interested in determining ways to improve the efficiency of missionary work in a specific developing country. To address this issue, one might conduct a study in which data are collected over a one-year time span for all of the Catholic missions in that country that have been active for at least ten years. At the end of the year it is determined whether each mission (a) showed an increase in the total number of registered members; (b) had any registered members enter into professional religion (class variables); or (c) had any charitable donations. Attributes that might be investigated include the length and/or orientation (pro-goodness or anti-sin) of the stereotypic sermon; psychoethnographic characteristics of the local constituency; type and quantity (measured in terms of either hours worked per week or percentage of theoretical need of one's constituency) of community service, educational programs, or extracurricular activities; availability of medical expertise, food, water, clothing, and temporary shelter; use of a direct versus an indirect recruitment approach; or the reverence of the community for traditional (indigenous) values and customs. ODA could be used to determine the optimal extent to which each attribute predicts whether or not a mission's membership increases, any of its members enter the religious service, or a charitable donation is received. Were data available concerning the actual change in the number of registered members, the actual number of registered members entering religious service, or the dollar (or other monetary index) amount of the charitable donation(s), ODA could determine the optimal ability of each attribute to predict the number of new registered members, the number of members entering religious service, or the monetary amount of charitable donations. Finally, were like data collected for missions representing other religious orientations, ODA could determine if findings generalized across religion, and, if findings did not generalize, ODA could determine religions for which consistent findings emerged.

Personnel Selection

Imagine that one is interested in using a questionnaire measure of motivation to predict whether applicants for a sales job would, if hired, be rated as being desirable or undesirable employees by their supervisors (class variable). To address this issue, one might conduct a study in which data are collected from 100 observations randomly selected from the sample of all applicants hired for a sales job under a one-year contract. At the end of the year, supervisors rate the desirability (pro or con) of the observations. With these data, ODA could determine the value of the score on the motivation test that resulted in maximum PAC when used to predict desirability ratings and could determine the value of the score that optimally predicted desirable (or undesirable) employees. Were ratings of observations made on more sensitive scales, one could use the actual ratings as weights and ODA would find the value of the score that maximized prediction of overall desirability ratings. Were objective measures of the profitability of the observations available, ODA could determine the value of the score that facilitated optimal prediction of the return of the hiring decisions. Scores on other questionnaires that assess additional factors that are theoretically relevant to job performance could also be investigated. Had data been collected for multiple samples, such as reflected by applicants for other types of jobs, ODA could determine the consistency of the findings across samples, and could identify samples for which findings were consistent. A particularly compelling example of the need for simultaneous multiple sample analysis is reflected in the legislation that regulates personnel selection practices. That is, the Equal Employment Opportunity Commission guidelines require that the relationship between scores on the test and supervisor-rated job performance should be consistent across samples created by crossing the factors of the observation's gender, age, and race. Using ODA, this would simply entail determining whether the model relating the score on the test to the rated desirability of the employee was consistent across the samples created by crossing gender, age, and race.

Prospecting

Is the presence of conduits from source to reservoir rock positively related to the presence of oil? To answer this question one might conduct a study in which an oil firm drilled 10 wells in scattered sites that all had conduits (cracks), and drilled 10 wells in close proximity to these but in areas without cracks. The presence versus absence of oil is recorded for each well. With these data, ODA can determine whether the presence of cracks is a reliable indicator of the presence of oil. If the number and/or size of the cracks were recorded, ODA could be used to determine the number and/or size of cracks that best predicts the presence versus absence of oil. Were the amount of oil recorded, ODA could determine the number and/or size of cracks that best predicts the amount of oil located. Were the return (i.e., the price received for oil minus price of getting the oil to market) recorded, ODA could determine the number and/or size of cracks that best predicts the return from one's oil prospecting. Other variables, such as the presence (versus absence), size, and number of reservoirs (sandstone, broken limestone), source rock (shale), or shale maturity (thermal alteration) could also be studied as possible predictors. Multisample analysis could be used to determine if the

ODA model for the prediction of oil generalizes to the prediction of natural gas or coal. Conceptually similar procedures could be used in other forms of prospecting, whether for commodities (water, mushrooms, medicinal plants, pearls), gems (rubies, diamonds, sapphires), metals (gold, silver, platinum, copper, uranium), or treasure (sunken or buried).

Selling Shoes

Imagine that one is interested in determining whether interpersonal style influences success at selling shoes. To answer this question, one might conduct a study involving all customers (observations) attended by a single salesperson during one month. Observations are randomly assigned to one of two conditions: the interpersonal style of the salesperson is either abrupt and directive, or cordial and receptive (attribute). If a record was kept of which observations bought shoes and which observations did not (class variable), ODA could determine the interpersonal style that best predicted the number of shoes sold. Were records kept of the price of (or profit from) the shoes sold, ODA could determine the interpersonal style that best predicted the value (profitability) of the shoes sold. In addition to the interpersonal styles described above, other perhaps more appropriate or refined interpersonal styles could be added to the design, and additional attributes (e.g., amount of time spent with the observation; the gender, age, or mood of the observation; the number of different shoe styles from which to choose) could also be evaluated. ODA could also be used to evaluate the generalizability of these findings across different salespeople.

Speeding

Imagine that a motorist is interested in discovering factors that predict whether one receives a speeding ticket on the local interstate (class variable). To address this issue, one might conduct a study in which data are collected for all the motorist's journeys on the interstate over the past year. The attribute of primary importance is velocity (miles per hour). With these data, ODA can determine the velocity that provides optimal predictability concerning whether or not one will receive a ticket, or can determine the maximum velocity that minimizes the number of tickets received. Were information recorded concerning the dollar amount of the fine for each ticket, ODA could determine the velocity that optimally predicts the amount of fines that one receives, or that minimizes the amount of the fines that one receives. Other attributes that might be investigated include, for example, the time of day; weather conditions (wind, temperature, precipitation); the presence versus absence of salient social events (holiday, big game, town meeting, riots); or the terrain (straight or curvy; flat or hilly; open or forested). Were data collected for multiple samples, such as different types or models of automobiles, or different interstates, ODA could determine the extent to which the findings generalized across samples and could identify samples for which consistent findings emerged. It seems worthwhile to note that the highway police probably know about this by now, and may be using the same analyses to predict how best to nab speeders! However, unlike individual motorists who are probably most interested in minimizing their personal speeding fines, the police may weight the ODA analyses by the number of accidents, injuries, or fatalities incurred, and thus attend to different critical velocities, locations, terrains, and so forth in their decision making.

Suicide

Imagine that one is interested in predicting with maximum PAC the nature of depressed people who attempt to commit suicide. To address this issue one might conduct a study in which data are collected for a five-year comprehensive consecutive sample of clinical inpatients (observations) in a psychiatric hospital. By order of the court, all observations are subject to an involuntary 21-day commitment for psychiatric observation, and all are diagnosed as depressed. Observations who attempt to commit suicide during their stay at the hospital are to be discriminated against observations that do not attempt suicide prior to being discharged. The attribute of primary interest is the score that observations achieve on an objective component of an intake interview, that reflects the degree to which an observation manifests depressive and suicidal ideation. With these data, ODA can determine a model—that is, a cutpoint value on the score—that optimally classifies observations who do versus do not attempt suicide. Also, ODA can determine a model that optimally classifies the observations that specifically attempt (or specifically do not attempt) suicide. The observations who succeeded in committing suicide might well be separated from those who attempted but failed, forming a third category. The optimal ability of other attributes—such as prior behavior, religiosity, drugs, parental status, or stressful life events—to predict suicide status could also be determined. Were data available, ODA could also be used to evaluate the generalizability of the findings across different clinics and to identify clinics with consistent findings.

Target Recognition

Imagine that one is interested in developing a methodology of screening for training as potential helicopter gunners individuals who will be best able to discriminate friendly from hostile tanks during close-encounter, congested, daylight desert warfare. In pursuit of this objective, one might conduct a study in which data are collected for a random sample of gunners with an outstanding record during close-encounter, congested, daylight desert warfare in the war against Iraq (i.e., who destroyed at least one enemy tank and no friendly tanks), and from a random sample of gunners with a horrendous record (i.e., who destroyed at least one friendly tank and no enemy tanks). Images of enemy and friendly tanks are individually presented to each observation in a helicopter warfare simulator, and the total number of seconds taken by each observation to maneuver gunnery radar to orient with an aggressive disposition toward all of the enemy (attribute 1) and friendly (attribute 2) tank stimuli is recorded. With these data, ODA could find optimal cutpoint values (threshold number of seconds) on each attribute for best discriminating good gunners versus bad gunners. Were the number of enemy (or friendly) tanks destroyed by observations during the war against Iraq recorded, ODA could provide a cutpoint (on both attributes) that resulted in optimal prediction of the number of enemy (or friendly) tanks destroyed. The number of enemy (friendly) tanks destroyed during the simulated battle could be used similarly, although the validity of actual warfare data may be greater. The ability of additional attributes, such as visual acuity, visual-motor coordination, or experience with video games to predict skill could also be investigated. Were data collected for multiple samples, such as for dawn, twilight, and night fighting, or for other types of targets (vehicles, dwellings, encampments) or aircraft (other helicopters, airplanes, jets), ODA could determine whether findings were consistent across samples and, if that were not the case, could identify samples for which consistent findings emerged. Were it desirable to minimize civilian carnage, similar methods could be used to train gunners to recognize and avoid firing on civilian vehicles, dwellings, or encampments. Conceptually similar methods may be used to understand and perhaps enhance the discriminatory accuracy of players, coaches, or referees in sports such as baseball (strike versus ball, safe versus out) or tennis (fair versus foul shot).

Teaching

Imagine that one is interested in improving the quality of elementary school education. To address this issue, one might conduct a study in which data are collected for all first grade students who are randomly assigned to one of 16 different elementary schools (matched on socioethnographic factors including intelligence; ethnicity; socioeconomic level, gender, and transportation time) in a single county. Of these schools, 8 emphasize traditional teaching methods, and 8 emphasize self-directed, self-paced teaching methods (class variable). At the end of the school year, students are tested on a standardized measure of general achievement (attribute). With these data, ODA could be used to determine the optimal extent to which the two teaching methods result in discriminable scores on the achievement test. Other attributes that might be predicted by teaching method and that could be investigated include, for example student grade-point averages; relative frequency of students graduating to the next grade level or graduating from college (longitudinal research); ratings on Likert-type scales of how much students enjoy school; the number of hours per day that students spend studying and/or watching television, and/or the type of television programming (including both educational and non-educational programs) watched; the amount of time spent listening to music, watching videos, playing video games or sports; the education level and/or occupation of parents, siblings, and other relatives and/or friends; or reinforcement contingencies at home. The socioethnographic variables, alone or in combination, constitute possible alternative class variables. Alternatively, the socioethnographic variables may be treated as multiple samples (as could data from grades 2–8, or from multiple counties or states): ODA could then determine whether findings were consistent across the samples, and, if results were inconsistent, could find samples for which consistent findings emerged.

Vacationing

Imagine that one is interested in identifying attributes that optimally facilitate the personal a priori knowledge of whether a planned vacation will be enjoyable. In order to develop a personal ODA capable of providing this knowledge, one might conduct a study in which data are collected for one's last ten vacations (see the history example). On the first day after one's return home from each vacation, a rating indicating whether the vacation was enjoyable is recorded (class variable). Possible attributes include, for example, the presence and/or number of others and one's relationship with them; weather conditions; the location and expense of the vacation; frequency of opportunity and types of activities; and one's health status. With these data, ODA could determine the optimal ability of each attribute to predict with maximum PAC whether the vacation was rated as being enjoyable. If subjective weights reflecting a more precise rating of the enjoyment experienced on vacation were made (e.g., 10-point Likert-type ratings), ODA could predict with maximum PAC the overall enjoyability of the vacations. A variation on this involves a travel agent coaching clients in this methodology in an attempt to optimize the enjoyability of their vacations. Post-vacation interviews and records of return business would be useful in validational analysis.

Weight Loss

How does one identify variables that predict with maximum PAC whether a person attempting to lose weight is successful? To address this issue, one might conduct a study in which data are collected from a sample of people who enroll in a weight loss course with the objective of losing weight. On the first day of the course, observations are weighed and complete a measure of self-efficacy (belief that "I can"). At the end of the course, it is determined whether observations have lost weight relative to their weight at the beginning of the course (class variable). With these data, ODA could determine the value of the score on the self-efficacy scale that predicted with maximum PAC whether observations lost weight. Were information recorded concerning the absolute amount of weight lost or gained (this enforces that individuals with the greatest absolute weight changes are weighted most strongly by the model), ODA could determine the value of the score on self-efficacy that predicted with maximum PAC the amount of weight lost or gained. The ability of additional attributes, such as gender, age, marital status, amount of daily free time, desk-bound or mobile occupation, home location (transportation time from parks, beaches, bike paths, or other recreational resources), or car ownership status to predict frequency or amount of weight loss success could also be investigated. Were data collected from other samples (such as other weight-loss groups; groups of individuals attempting to terminate smoking, drinking, or other substance abuse behavior; or groups undergoing different clinical intervention methods), ODA could be used to assess whether findings were consistent across samples and to identify samples with consistent findings.

Zoology

What factors are related to the birth (class variable) of lion cubs in captivity? To address this question, one might conduct a study in which data from all U.S. zoos with at least one pair of sexually mature male and female lions are collected. For every non-sterile sexually mature female lion, it is determined whether at least one surviving cub was born (class variable) during 1992. The attribute is the length of time that the female had been in captivity (attribute). With these data, ODA could determine the length of captivity that optimally predicted whether the lioness would bear a cub. Were the number (or overall weight) of cubs recorded, ODA could determine the length of captivity that predicted with maximum PAC the number (weight) of cubs born. Other attributes that might predict the fertility of lions that could be investigated, for example, might include the number of human visitors; distance from visitors to lions; mean noise level; mean temperature, humidity, or barometric pressure; the age of the female or the male; or the length of time the female and male lions had cohabited. Using ODA, the generalizability of the findings could be assessed over multiple samples, such as data collected for other felines (tigers, pumas, bobcats), mammals (wolves, polar bears, elephants), or animals (reptiles, birds, fish). If findings were inconsistent across all samples, ODA could be used to identify the samples for which consistent findings emerged.

Now It Is Time to Start Analyzing Data

There is a new tool on the block. Applied research in a myriad of substantive areas has been conducted over the past decade using earlier versions of the software accompanying this book, and has met with stellar reception vis-à-vis publication in a host of scientific journals. As we hope the hypothetical examples illustrated, the ODA paradigm applies optimally to any quantitative data set and can be specifically tailored to evaluate any stated hypothesis—truly a "designer" statistical methodology. At this point in the discussion, we believe that many readers must be eager to witness the new tool in action. That is what we turn to next.

CHAPTER 2

Using the ODA Software

To begin, create a directory on your hard drive, into which you will copy the ODA software system. We name our directory "ODA" on our computer. Once the directory is created, copy everything on the CD to the directory. The ODA system is now ready for action.

In this chapter we begin by describing the ODA commands and then we illustrate how to run the ODA software using Programmer's File Editor (PFE), an intuitive, easy-to-use integrated editing application for text files, created by Alan Phillips.

The ODA Commands

The ODA system contains a flexible scripting language, which enables the user to specify precisely the nature of the analysis. A myriad of experimental structures may be defined by using combinations of its commands. The ODA system can be used to analyze problems involving as many as 500 variables, 65,536 observations, and 16 groups (to understand what groups are in the ODA system requires knowledge about the Gen feature). The following is an alphabetical listing of these commands, along with explanations of their associated keywords. At this point, these commands may make little sense to you. The commands will become childishly simple once you have completed this book.

ATTRIBUTE

    Syntax ATTRIBUTE *variable list* ;
    Alias ATTR
    Remarks The ATTRIBUTE command specifies the attribute variable or variables to be used in the analysis. This command is required unless TABLE input is specified. If more than one variable is named, a separate analysis will be run for each

variable. The TO keyword may be used to define multiple variables in the variable list. For example, the command

    ATTRIBUTE A1 TO A4 ;

indicates that the variables A1, A2, A3, A4 will be treated as attributes. Further exposition of the TO keyword may be found in the discussion for VARS.

CATEGORICAL

Syntax CATEGORICAL {ON | OFF} ;
CATEGORICAL *variable list* ;

Alias CAT

Remarks The CATEGORICAL command specifies that categorical analysis will be used: It is required when the attribute to be analyzed is categorical. Using the ON keyword indicates that all variables in the variable list are categorical. CATEGORICAL with no parameters is the same as CATEGORICAL ON. The TO keyword may be used in the variable list (see the discussion under VARS). When using TABLE input, a CATEGORICAL analysis is assumed, and it is not necessary to specify this command.

CLASS

Syntax CLASS *variable list* ;
CLASS {ROW | COL} ;

Remarks The CLASS command specifies the class variable to be used in the analysis. This command is mandatory. If more than one variable is named, a separate analysis will be run for each variable. ROW and COL are used for TABLE input to indicate whether the rows or columns of the table are to be used for the class variable. Otherwise, the TO keyword may be used in the variable list (see the discussion under VARS).

DATA

Syntax DATA ;

Remarks The DATA command indicates that the entries that follow the command are data to be used in the current analysis. The END statement terminates the data block. For example, the following commands enter hypothetical data on two variables for each of two observations:

    DATA;
    1 2
    3 4
    END;

Each line should correspond to a single observation. The END (or END DATA) statement must be the only command on the line on which it appears.

DEGEN

Syntax DEGEN {ON | OFF} ;
DEGEN *variable list* ;
Alias DEGENERATE
Remarks DEGEN specifies whether degenerate cutpoints are allowed. If DEGEN OFF is specified, the resulting ODA solution must have at least one observation assigned to each predicted class. DEGEN allows flexibility in data sets which have small or no representation in some classes. The default is OFF. DEGEN with no parameters is the same as DEGEN ON. The TO keyword may be used in the variable list (see the discussion under VARS).

DIRECTION

Syntax DIRECTION { < | LT | > | GT | OFF} *value list* ;
Aliases DIR, DIRECTIONAL
Remarks The DIRECTION command specifies the presence and nature of a directional (i.e., an a priori or one-tailed) hypothesis. The parameter < or LT indicates that the class values in the value list are ordered in the "less-than" direction. The parameter > or GT indicates the class values are ordered in the "greater-than" direction. The value list must contain every value of the class variable currently defined. The default is OFF.

EXCLUDE

Syntax EXCLUDE *variable* {= | <> | < | > | <= | >= | OFF} *value* (,*value2*,...) ... ;
Aliases EX, EXCL
Remarks This command excludes observations with the indicated *value* of *variable*. For example,

EXCLUDE C=3 ;

tells ODA to drop all observations with the value of 3 for variable C. Also, the command

EXCLUDE A=1 G>=902 ;

drops all observations with the values of 1 for variable A or values greater than or equal to 902 for variable G. Commas in the exclude string enable the user to exclude multiple values of a variable with a single command:

EXCLUDE B=1,3 ;

excludes all observations which have a value of 1 or 3 for variable B. Multiple EXCLUDE commands may be entered, up to a maximum of 100 clauses. The system will exclude observations that satisfy any of the EXCLUDE clauses. EXCLUDE is not allowed with TABLE input.

FREE

See VARS.

GO

Syntax GO ;
Remarks The GO command begins execution of the currently defined analysis.

GEN

Syntax GEN {OFF} *variable* ;
GEN TABLE *g* ;
Alias GROUP
Remarks The GEN command specifies the variable whose (integer) values indicate groups in a multisample (GEN) analysis. If TABLE has been specified, then GEN TABLE *g* indicates that *g* tables are present, corresponding to *g* GEN groups. The default is OFF.

HOLDOUT

Syntax HOLDOUT *path\file name* ;
Alias HOLD
Remarks HOLDOUT specifies the file name to be used for hold-out (validity) analysis. The variable list for the hold-out file must be in the same order as that for the main input file. OFF turns the hold-out indicator off.

ID

Syntax ID {OFF} *variable* ;
Remarks The ID command defines the ID variable that is to be printed in the long report. The default is OFF.

INCLUDE

Syntax INCLUDE *variable* {= | <> | < | > | <= | >= | OFF} *value* (*,value2,...*) ... ;
Aliases IN, INCL

Remarks The INCLUDE command functions in the same way as the EXCLUDE command, except that ODA will keep only those observations with the indicated *value* for *variable*. If multiple INCLUDE statements exist, only those observations will be kept which satisfy all these INCLUDE statements. INCLUDE is not allowed with TABLE input.

LOO

Syntax LOO {ON | OFF} ;

Remarks The LOO command specifies that a leave-one-out (jackknife) analysis will be performed. LOO is not allowed in WEIGHTed CATEGORICAL problems. The default is OFF. LOO with no parameters is the same as LOO ON.

MCARLO

Syntax MCARLO {ITERATIONS *value* | SECONDS *value* | TARGET *value* | SIDAK *value* | STOP *value* | STOPUP *value* | ADJUST | OFF} ;

Alias MC

Remarks The MCARLO command controls Monte Carlo (simulation) analysis for estimating Type I error, or p. The keywords specify a number of stopping criteria; if any criterion is met, then the analysis stops. ITERATIONS (ITER) specifies the maximum number of Monte Carlo iterations, SECONDS (SEC) specifies the maximum number of seconds before the analysis terminates. TARGET specifies a target significance level. SIDAK adjusts the target to reflect a Sidak (Bonferroni) adjustment of the TARGET level, in which *value* is an integer that indicates the number of experiments involved in the adjustment (see chap. 3, this volume). STOP indicates the confidence level (in percent) that the estimated Type I error rate is *less* than the TARGET value, at which point the analysis stops. STOPUP indicates the confidence level (in percent) that the estimated Type I error rate is *greater* than the TARGET value, at which point the analysis stops. For example, the command

MCARLO ITER 1000 SEC 30 TARGET .01 STOP 99.9 STOPUP 99 ;

indicates that a Monte Carlo analysis will be conducted, and will stop when one of the following occurs: (1) 1,000 iterations have been executed, (2) 30 seconds have elapsed, (3) a confidence level of 99.9% has been obtained for $p < 0.01$, or (4) a confidence level of 99% has been obtained for $p > 0.01$. The default Monte Carlo method is conservative in the estimation of significance, in that a Monte Carlo iteration, whose optimal value is tied with the optimal value obtained from the original analysis, is always counted toward a higher significance level. Specifying

ADJUST will adjust for this boundary by splitting these tied iterations in half. The default is OFF.

MISSING

Syntax MISSING {*variable list* | ALL} (*value*) ;
Alias MISS
Remarks The MISSING command tells ODA to treat observations with value (*value*) as missing for each variable on the list. For example, the command
    MISSING A B C (–1) ;
indicates that observations with variables A, B, or C equal to –1 will be dropped if they are present in a CLASS, ATTRIBUTE, WEIGHT, GROUP, or ID variable. ALL specifies that the indicated missing value applies to all variables. The TO keyword may be used in the variable list (see the discussion under VARS).

OPEN

Syntax OPEN {*path\file name* | DATA} ;
Remarks The OPEN command specifies the data file to be processed by ODA. This file must be in ASCII format. DATA indicates that a DATA statement, with inline data following, appears in the command stream.

OUTPUT

Syntax OUTPUT *path\file name* {APPEND} ;
Remarks The OUTPUT command specifies the output file containing the results of the ODA run. The default is ODA.OUT. APPEND indicates that the report is to be appended to the end of an already existing output file.

PRIMARY

Syntax PRIMARY {MAXSENS | MEANSENS | SAMPLEREP | BALANCED | SENS *value* | DISTANCE | RANDOM | GENMEAN | GENSENS *value* | DEFAULT} ;
Alias PRI
Remarks The PRIMARY command specifies the primary criterion for choosing among multiple optimal solutions. MAXSENS (or MAXPAC) is maximum sensitivity. MEANSENS (MEANPAC) is the mean of the sensitivities of the separate classes. SAMPLEREP (SREP) selects the pattern of predicted class membership most closely resembling the sample class membership. BALANCED (BAL) selects the solution in which the sensitivity of the actual classes is most similar amongst each

other. SENS (PAC) selects the solution with maximum sensitivity of class *value*. DISTANCE (DIST) selects the solution with smallest maximum (over all cutpoints) distance between the cutpoints and their boundaries. RANDOM (RAND) selects a randomly chosen solution. GENMEAN is used only when GEN is in effect. It selects the solution with maximum mean (weighted) sensitivity over all GEN groups. GENSENS selects the solution with the maximum (weighted) sensitivity of group *value*. The default is MAXSENS when PRIORS is ON and MEANSENS otherwise.

PRIORS

Syntax PRIORS {ON | OFF} ;
Remarks The PRIORS command indicates whether the ODA criterion will be weighted by the reciprocal of sample class membership. The default is ON. PRIORS with no parameters is the same as PRIORS ON.

QUIT

Syntax QUIT ;
Remarks Use the QUIT command to exit from ODA immediately.

REPORT

Syntax REPORT {SHORT | LONG} ;
Alias REP
Remarks The REPORT command specifies whether the short or long report is to be generated. The LONG report additionally prints the predicted and actual class memberships for each observation (ordered analysis) or for each cell (categorical analysis). The default is SHORT.

RESET

Syntax RESET ;
Remarks Use this command to reset all parameters to their default values.

SECONDARY

Syntax SECONDARY {MAXSENS | MEANSENS | SAMPLEREP | BALANCED | SENS *value* | DISTANCE | RANDOM | GENMEAN | GENSENS *value* | DEFAULT} ;
Alias SEC

Remarks The SECONDARY command specifies the secondary criterion for choosing among multiple optimal solutions. The default is SAMPLEREP. See the entry for PRIMARY for definitions of the above criteria.

SEED

Syntax SEED {*value* | TIME | 0} ;

Remarks The SEED command supplies the seed value for random number generation. TIME or 0 indicate that the current time will be used for the seed. If this command is not present, the time at program initiation will be used.

TABLE

Syntax TABLE *row* (*col*) ;

Alias FREE TABLE

Remarks The TABLE command is used for categorical analysis only, and indicates that a *row*-by-*col* table is present in the input file. If only *row* is entered, a square *row*-by-*row* table is assumed. For the sake of illustration, imagine that the following 2-by-2 table constitutes the data one wishes to analyze:

	Column 1	Column 2
Row 1	5	6
Row 2	7	8

In the ODA script for this illustration, the statement TABLE 2 would be used to indicate that the 2-by-2 table was to be input. CLASS ROW would indicate that the rows were to be considered the class variable. If the table is rectangular, the CLASS command should reflect the smaller value of *row* or *col*. For example, if

TABLE 4 3 ;

was entered, the user should then enter

CLASS COL ; .

When using TABLE input, a CATEGORICAL analysis is assumed, and it is not necessary to specify this command.

TITLE

Syntax TITLE *title* ;

Remarks The TITLE command specifies the title to be printed in the report. TITLE with no parameters erases the currently defined title.

VARS

Syntax VARS *variable list* ;

Alias FREE

Remarks The VARS command specifies a list of variable names corresponding to fields in the input data set. The TO keyword may be used to define multiple variables in the variable list. For example, the command
VARS A B C X1 TO X5 ;
specifies that the input file contains, in order, variables A, B, C, X1, X2, X3, X4, and X5, and that there is at least one blank space separating all adjacent data. Alternatively, the data points may be separated by a single comma or tab (with no spaces).

The TO keyword may only be used to input a range of variables that have the same name except for the integer at the end of the name. The integers must be positive and ascending, increasing one unit per variable. Thus, VAR1 TO VAR10 is admissible (defining 10 variables). In contrast, VAR10 TO VAR1, VARA TO VARJ, or A TO X10, are not admissible.

The data for each observation may all exist on a single line of the data set, or may be spread on multiple adjacent lines. It is not recommended that a new observation be included on a line that contains data from the previous observation.

WEIGHT

Syntax WEIGHT {*variable* | OFF} ;
Alias RETURN
Remarks The optional WEIGHT command specifies the weight variable for the analysis. The data values for the WEIGHT variable supply the weight for the corresponding observation. The default is OFF.

Running the ODA Software

There are basically two methods for running ODA software. The first method involves the "Programmer's File Editor" (PFE), which allows an operator to edit an unlimited number of files in separate windows. The second method involves using the "old-fashioned" command line editor in the MS-DOS command prompt window. To understand this procedure, please refer to your Windows documentation. When in the MS-DOS command prompt window, execute the ODA program using the command: ODA *filename*. For example, if one's ODA program is called "ex51.cmd", enter: ODA ex51.cmd.

Which system one uses is a matter of personal preference because both ultimately get the job done. In the case of the authors, for example, one prefers to use the command prompt window, whereas the other prefers to use PFE, which we describe below.

Using ODA With PFE

Programmer's File Editor (PFE) is an intuitive, easy-to-use integrated editing application for text files. It was created by Alan Phillips. With PFE one can write and edit ODA scripts and data files, execute analyses, and view outputs from multiple runs. To illustrate the use of PFE, we will run an example from chapter 5.

To begin, launch the PFE program. To do this, in Windows Explorer navigate to the directory into which you copied the CD containing the ODA system. For expository purposes, we will assume that the name of this directory is the ODA directory. Inside this directory is a PFE subdirectory. Double-click on the icon labeled "PFE32.EXE". Once the PFE screen appears, we will go and get an ODA program to run. This is accomplished by pointing the mouse pointer at the "open file" icon, as shown in Figure 2.1.

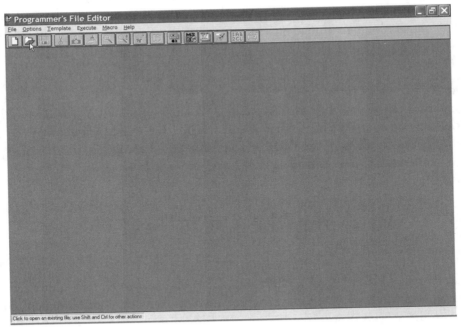

Figure 2.1.

(alternatively, choose Open from the File menu.) Next, using the PFE editor, navigate back to the ODA directory, place the mouse pointer on the "examples" subdirectory (see Figure 2.2) and click on it, and then click Open.

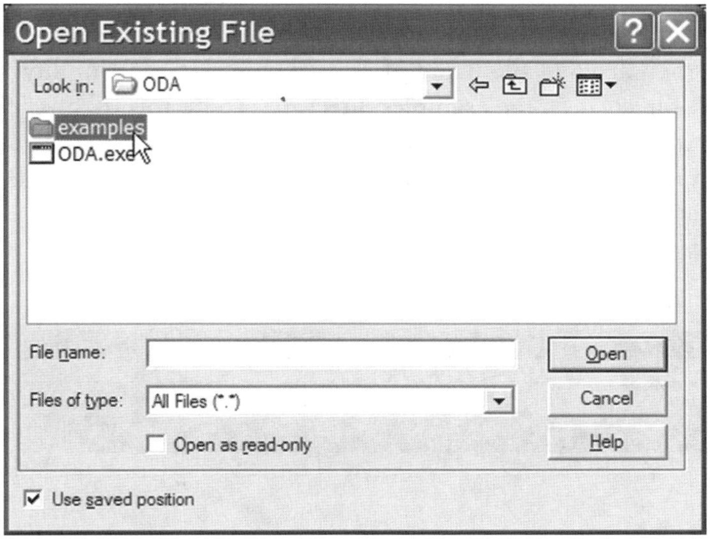

Using the ODA Software

Figure 2.2.

The files with the "cmd" suffix are executable ODA scripts, and the files with the "dat" suffix are data files (in your own work, of course, you may choose whatever suffix you wish. For example, one author uses the suffix "pgm" to identify ODA programs, and the other uses the suffix "x" to identify ODA programs). Click on "ex51.cmd" (see Figure 2.3), and then click Open.

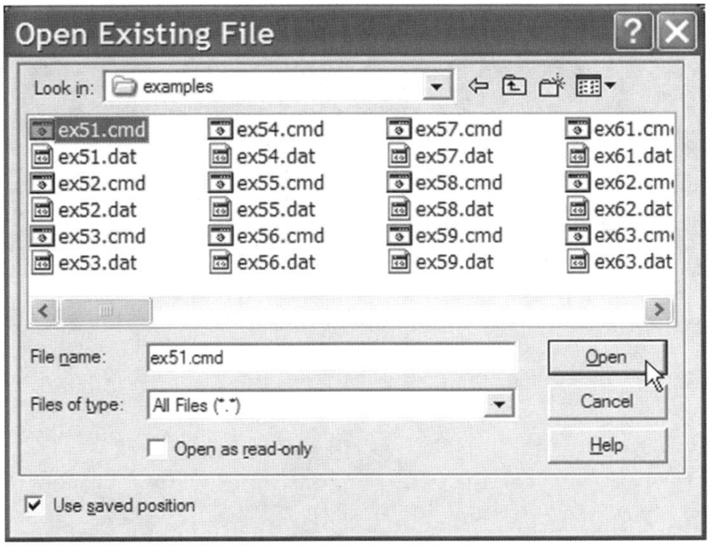

Figure 2.3.

The window that appears contains the ODA script for Example 5.1. The title appearing above the text is the path and file name of the ODA script. To simplify file management, change the default directory to the examples directory. To do this, right-click on the window (anywhere on the window) and choose "Change To File's Directory"—the next-to-last option in the pop-up window that appears when right-clicking in the window (see Figure 2.4).

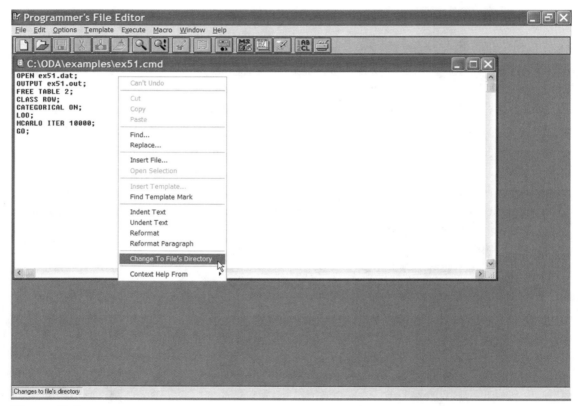

Figure 2.4.

Now it is time to execute the analysis. Click on the "Execute DOS Command" icon, which is the fourth icon from the right in the toolbar (see Figure 2.5).

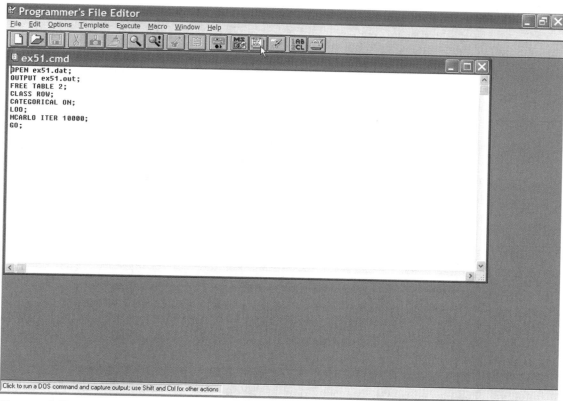

Figure 2.5.

After clicking on the icon, a "dialog window" will appear (see Figure 2.6). The dialog window contains two boxes, "Command" and "Directory". In the Command box, enter "oda.exe" along with the full path name, followed by "%p" (see Figure 2.6). In the Directory box, enter a period ("."). This indicates that the default directory will be used to look for the location of the files declared in the ODA script.

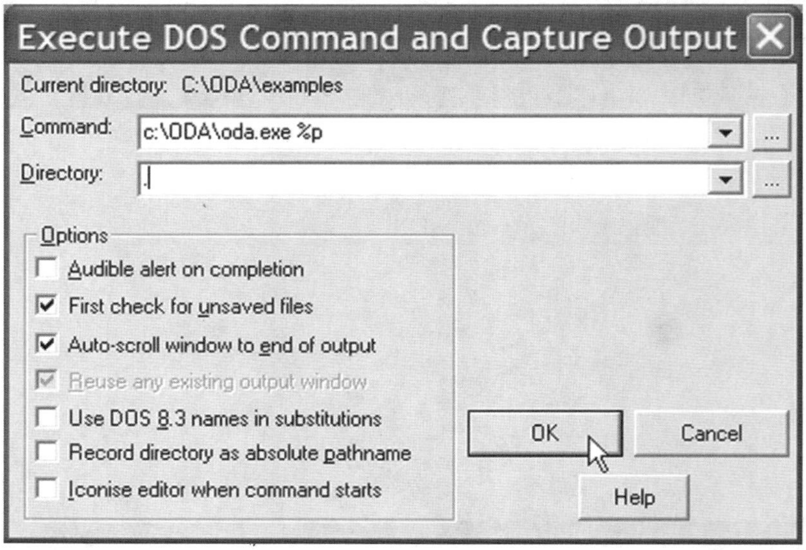

Figure 2.6.

Click OK to begin execution of the analysis. When this is done, the screen will appear as is shown in Figure 2.7.

Figure 2.7.

When the analysis is completed the DOS window will disappear, and a new window, called "Command Output" (with a number, 1, 2, etc., depending on how many DOS windows were opened), will appear. Now, navigate to the toolbar, and click the "Open" icon again (see Figure 2.1), and then choose the output file declared in the ODA script. In the present example the output file in the ODA script is called "ex51.out" (see Figure 2.5). So, as seen in Figure 2.8, click the mouse pointer on the file "ex51.out" (to highlight it), and then point and click on "Open."

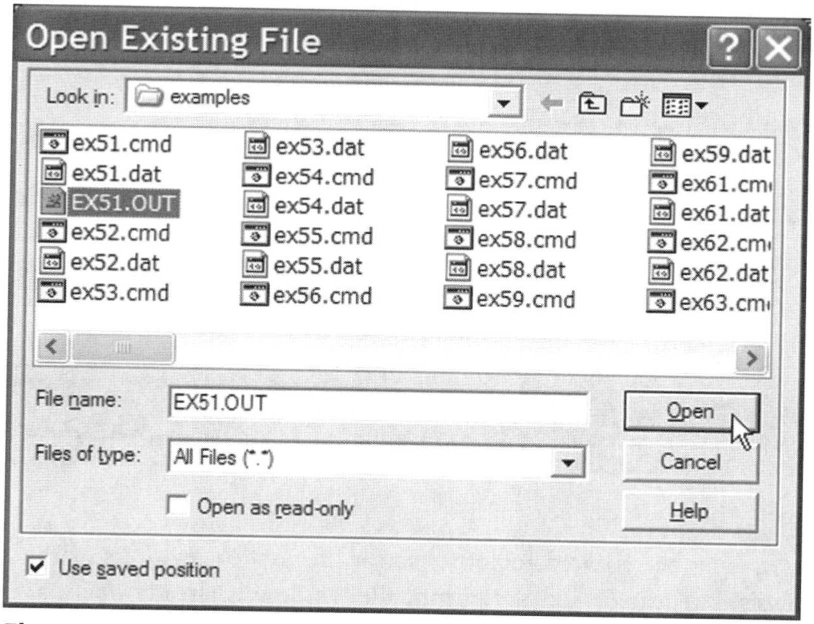

Figure 2.8.

The window displayed (Figure 2.9) contains the report from the ODA analysis.

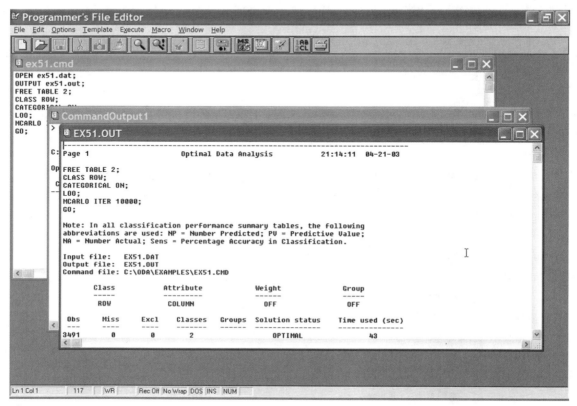

Figure 2.9.

This process may be repeated for other script files or for revisions of the current script file. If an analysis is repeated, and the output file window is already open, a dialog window will open up, asking whether the newly written output file should be displayed. Click "Yes" (see Figure 2.10). This will replace the output from the prior analysis with the output from the following analysis.

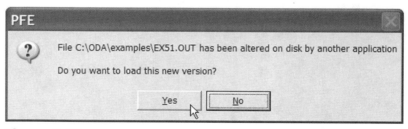

Figure 2.10.

Creating a Data Set for Analysis by ODA

The type of file that ODA software uses—ASCII—is among the most transportable formats across computers, and can easily be read by most commercial software. Here we describe how one can make a data set or ODA script file for analysis by ODA software.

Creating a Data Set or ODA Script File Using PFE

To create a data set or script file using PFE, click on the "new file" icon (Figure 2.11) or choose New from the File menu. Then type the data in this window (verify that the VARS command accurately reflects the order in which data are entered). Although a data set is illustrated, ODA scripts may be written in the same manner.

Figure 2.11.

Save the contents of the windows by clicking on the "save file as" icon as shown in Figure 2.12.

Figure 2.12.

Or, choose Save File As from the File menu and navigate to the desired folder (Figure 2.13).

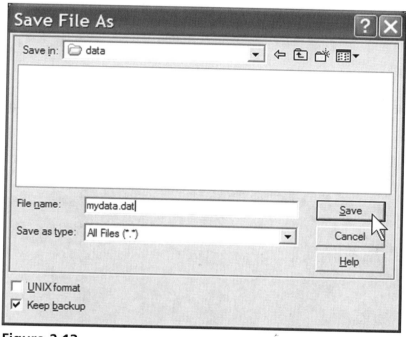

Figure 2.13.

Copying a Data Set From Excel

There are two methods for transferring data from Microsoft Excel to PFE, in order to be processed by ODA. The first is by using copy and paste. If the Excel worksheet has a header row, select all data except the header row—ODA can't read comments in a data file (Figure 2.14).

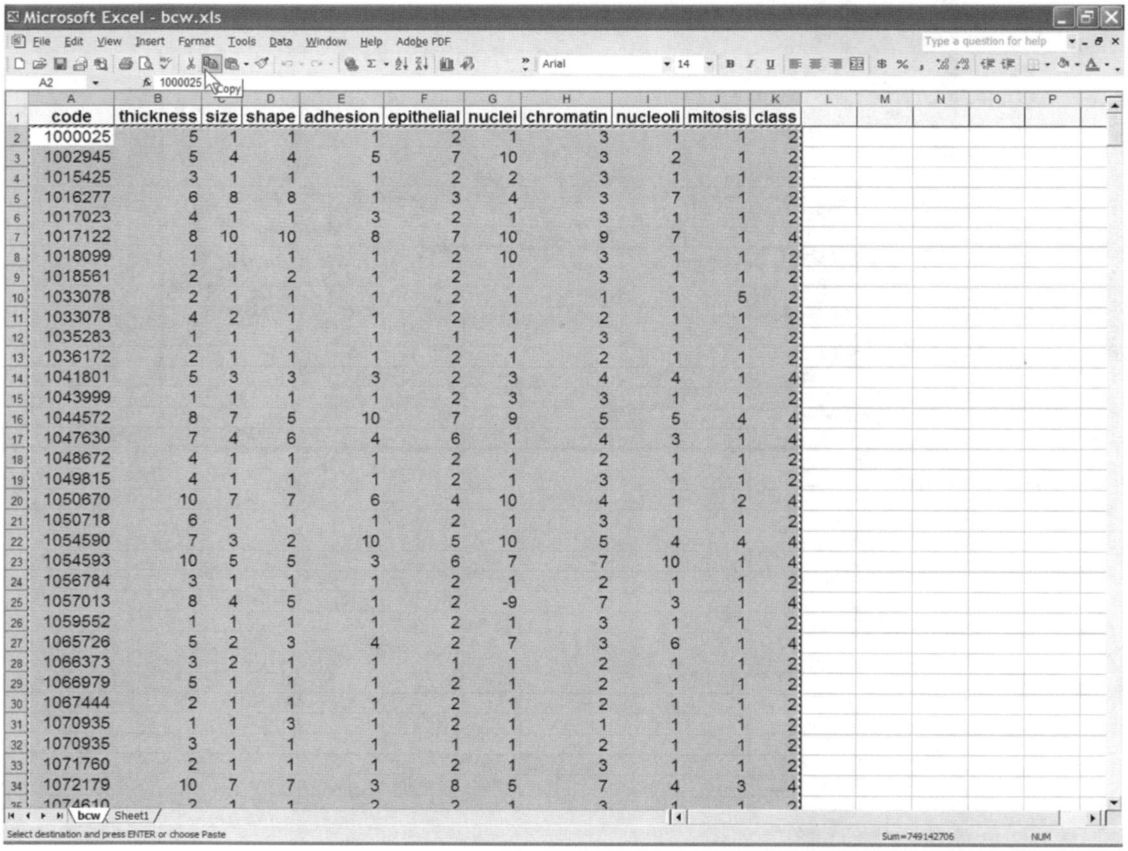

code	thickness	size	shape	adhesion	epithelial	nuclei	chromatin	nucleoli	mitosis	class
1000025	5	1	1	1	2	1	3	1	1	2
1002945	5	4	4	5	7	10	3	2	1	2
1015425	3	1	1	1	2	2	3	1	1	2
1016277	6	8	8	1	3	4	3	7	1	2
1017023	4	1	1	3	2	1	3	1	1	2
1017122	8	10	10	8	7	10	9	7	1	4
1018099	1	1	1	1	2	10	3	1	1	2
1018561	2	1	2	1	2	1	3	1	1	2
1033078	2	1	1	1	2	1	1	1	5	2
1033078	4	2	1	1	2	1	2	1	1	2
1035283	1	1	1	1	1	1	3	1	1	2
1036172	2	1	1	1	2	1	2	1	1	2
1041801	5	3	3	3	2	3	4	4	1	4
1043999	1	1	1	1	2	3	3	1	1	2
1044572	8	7	5	10	7	9	5	5	4	4
1047630	7	4	6	4	6	1	4	3	1	4
1048672	4	1	1	1	2	1	2	1	1	2
1049815	4	1	1	1	2	1	3	1	1	2
1050670	10	7	7	6	4	10	4	1	2	4
1050718	6	1	1	1	2	1	3	1	1	2
1054590	7	3	2	10	5	10	5	4	4	4
1054593	10	5	5	3	6	7	7	10	1	4
1056784	3	1	1	1	2	1	2	1	1	2
1057013	8	4	5	1	2	-9	7	3	1	4
1059552	1	1	1	1	2	1	3	1	1	2
1065726	5	2	3	4	2	7	3	6	1	4
1066373	3	2	1	1	1	1	2	1	1	2
1066979	5	1	1	1	2	1	2	1	1	2
1067444	2	1	1	1	2	1	2	1	1	2
1070935	1	1	3	1	2	1	1	1	1	2
1070935	3	1	1	1	1	1	2	1	1	2
1071760	2	1	1	1	2	1	3	1	1	2
1072179	10	7	7	3	8	5	7	4	3	4
1074610	2	1	1	2	2	1	3	1	1	2

Figure 2.14.

Then create a new window in PFE and click the "paste" icon or choose Paste from the Edit menu (Figure 2.15). The window can then be saved as a file as described above.

Figure 2.15.

If the Excel worksheet does not contain a header row, then it may be saved as an ASCII file directly from Excel. Choose Save As from the File menu in Excel, and select Text (Tab delimited) in the "Save as type" box, in order to save the worksheet as a tab-delimited file (Figure 2.16).

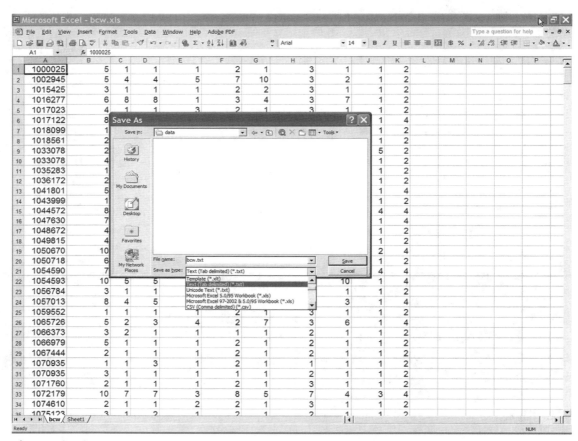

Figure 2.16.

The worksheet may also be saved as a comma-delimited file by selecting CSV (Comma delimited) in the "Save as type" box (Figure 2.17).

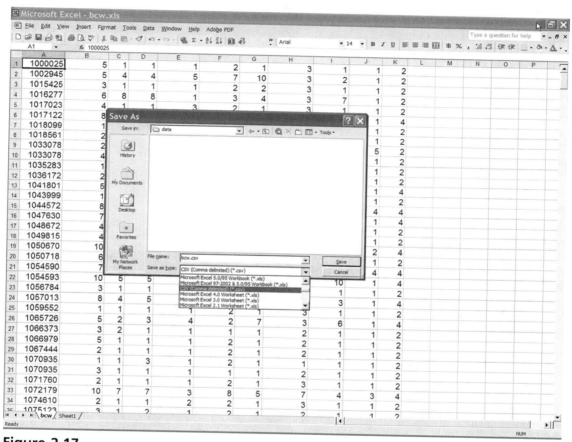

Figure 2.17.

The file is then ready to be opened in PFE (Figure 2.18).

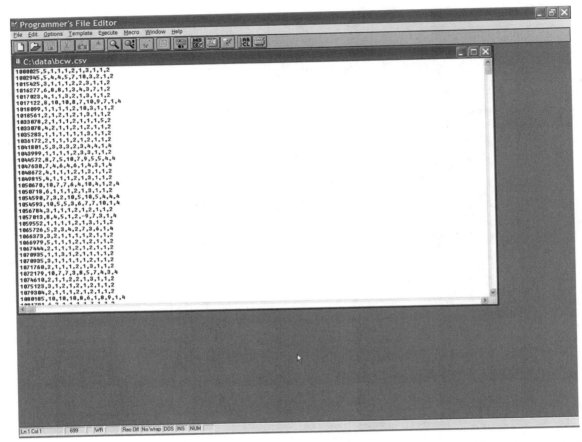

Figure 2.18.

Creating ASCII Files From SPSS

Saving an ASCII data set from an SPSS data set is similar to the above-mentioned Excel procedure. Simply open the SPSS data set and click on "Save As" under the file command. In the Save Data As window, click on the down arrow associated with the "Save as type" box, and select "Fixed ASCII (*.dat)". Now navigate to the folder and enter the filename in which the ASCII data will be saved, and click on Save (Figure 2.19).

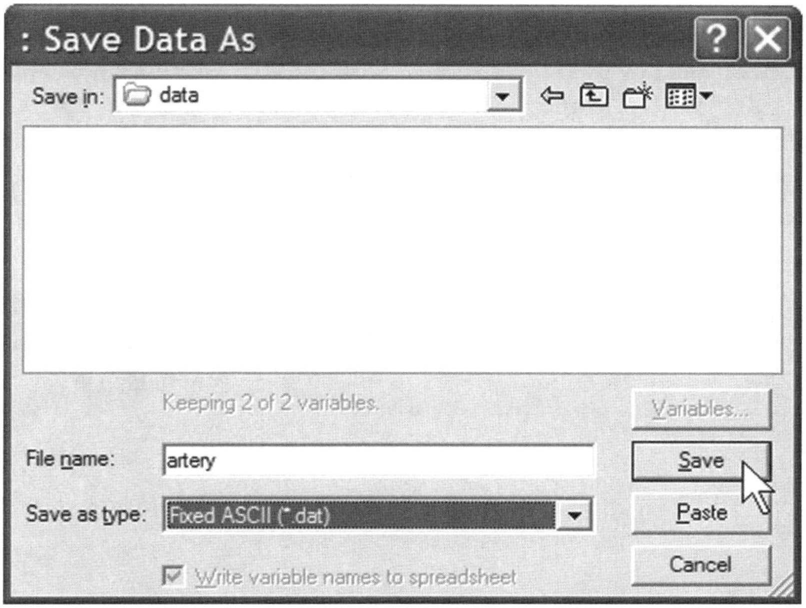

Using the ODA Software 53

Figure 2.19.

One also has the option of choosing only certain variables from a larger SPSS data set to save to an ASCII data set that requires fewer variables; this can be accomplished by clicking on the Variables command in the Save Data As window and selecting the variables that will be saved.

Alternatively, one may prefer to save the file in tab-delimited format. In the Save Data As window, click on the down arrow associated with the "Save as type" box, and select "Tab-delimited (*.dat)". Make sure that the box "Write variable names to spreadsheet" is *not* checked; typically, the default is that this box is checked, so click on the checkbox and it will disappear. Now navigate to the folder and enter the filename in which the ASCII data will be saved, and click on Save (Figure 2.20).

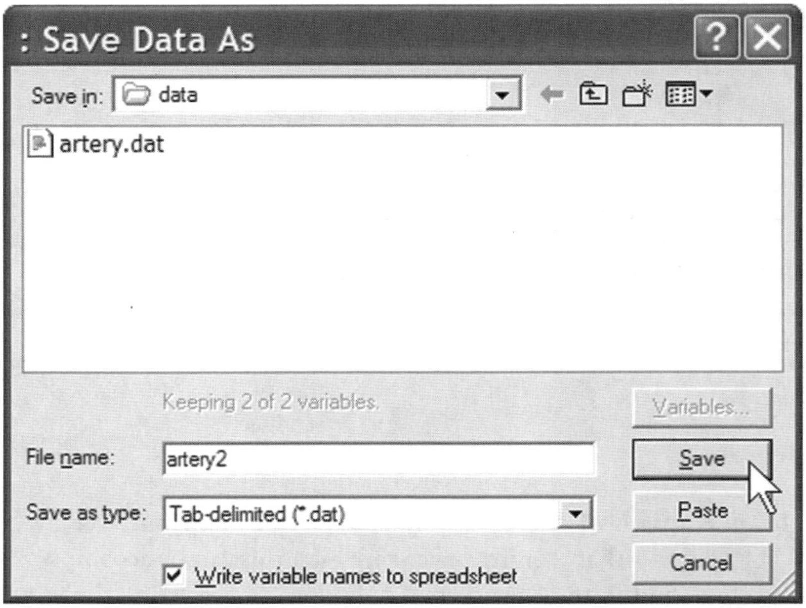

Figure 2.20.

Creating ASCII Files From SAS

Saving an ASCII data set from an SAS data set involves five commands. For example, imagine a hypothetical SAS data set ("DATA q;"), from which we wish to pull and write an ASCII data set containing three variables (e.g., sex, age, income). Further imagine we wish for the ASCII data set with the three variables to reside at and be called c:\research\ASCII.dat. SAS code required to accomplish this task is

```
DATA Q2;
SET Q;
FILE 'c:\research\ASCII.dat';
PUT sex age income;
RUN;
```

Missing Data

A very important point that one cannot overlook is that *all system missing data must be changed to a specified missing numeric value* prior to analysis via ODA. In our laboratory we use "–9" to represent missing data, if this value is not otherwise present in the data set. By using a single value such as –9 to identify missing data, one can specify all missing data for all variables in the ODA program using the command: "missing all (–9);".

It is tempting at this point, for us at least, to describe the nature of the ODA analysis output. As we discussed previously regarding the list of ODA commands, this would be premature. First it is necessary to learn the theoretical tenets of the ODA paradigm, step-by-step.

CHAPTER 3

Evaluating Classification Performance

The question of how best to summarize and interpret the classification performance achieved by a statistical model for a training sample has been widely discussed (Eisenbeis, 1977; Green, 1988; Hosmer & Lemeshow, 1980; Lachenbruch, 1975; Sorum, 1972). Growing consensus suggests that model performance appraisal based solely on statistical significance findings is inappropriate for evaluating classification and forecasting methods (Azar, 1997; Baumeister & Tice, 1996; Cohen, 1994; Goodman & Royall, 1988; Hagen, 1997), and that emphasis should instead focus on the ability of the methodology to achieve clinically (Feinstein, 1988; Kraemer, 1992) or otherwise context-specifically defined ecologically significant levels of predictive performance (Bacus & Gose, 1972; Eisenbeis, 1977; Nishikawa, Kubota, & Ooi, 1983; Yarnold, 1992).

Percentage accuracy in classification (PAC) is a widely cited and highly intuitive index of classification performance, but it is not the only way to conceptualize the accuracy of a classification method (Friedman, 1987; Kraemer, 1992; Rosner, 1982). Several additional indices of classification performance that are widely reported include sensitivity (the probability that an observation actually from class category c will be classified into c), and predictive value (the probability that an observation classified into c actually is a member of c). To facilitate clarity, imagine that the following classification results emerged in an ODA analysis:

Actual Class Membership	Predicted Class Membership		Row Marginal
	Class 0	Class 1	
Class 0	20	5	25
Class 1	10	15	25
Column Marginal	30	20	

In this so-called "confusion table," the actual status of the observation is given in the rows, and the predicted status of the observation is given in the columns. For example, looking across the top row of the table, we see that there were a total of 25 observations that were in actuality members of Class 0 (see row marginal column). Of these 25, a total of 20 were correctly predicted to be in Class 0, and a total of 5 were incorrectly predicted to be in Class 1. Looking across the bottom row of the table, we see that there were a total of 25 observations that were in actuality a member of Class 1 (see row marginal column). Of these 25, a total of 10 were incorrectly predicted to be in Class 0, and a total of 15 were correctly predicted to be in Class 1. Indices of model classification performance are based upon analysis of the pattern of actual class membership crossed by predicted class membership, as represented in the confusion table.

The first index of classification accuracy that is used to evaluate an ODA model is *overall classification accuracy* (or *overall PAC*), defined as 100% times the ratio of the total number of correct classifications divided by the total sample size. In numerical taxonomy this ratio is called the *simple matching coefficient* (Sokal & Sneath, 1963). In the hypothetical example, overall PAC = [(20 + 15) / 50] × 100% = 70%: Thus, 70% of the observations were correctly classified, and 100% − 70% = 30% of the observations were misclassified. Regardless of whether the application is *balanced* (i.e., has class categories having the same number of observations), *if no weights are specified* (i.e., one is interested only in maximizing overall PAC), then ODA *maximizes overall PAC*.

The second index of classification accuracy used to evaluate an ODA model is *sensitivity*, which indicates the percentage of the membership of each class category that was correctly classified. Sensitivity reflects the ability of the model to discriminate observations from different class categories. It is an index of the *descriptive utility* of an ODA model, reflecting its proficiency in characterizing different class categories. The sensitivity of the model for Class Category 1 gives the percentage of the actual Class 1 observations that were accurately classified. The sensitivity for Class Category 0 gives the percentage of the actual Class 0 observations that were accurately classified. In epidemiology this value is known as the model *specificity* (Friedman, 1987). For *multicategory* applications involving more than two class categories (C > 2), the sensitivity for class category c gives the percentage of the actual class c observations accurately classified. In the present example, sensitivity for Class 0 = (20 / 25) × 100% = 80%, and sensitivity for Class 1 = (15 / 25) × 100% = 60%. Thus, the model correctly classified 80% of the actual Class 0 observations, and 60% of the Class 1 observations. The *mean sensitivity across classes*—or *mean PAC*—is (80% + 60%) / 2 = 70%. *Weighting by prior odds maximizes the mean PAC yielded by an ODA model.* This is analogous to the use of *antecedent probability* or *base rate* in Fisher's discriminant analysis, and simply involves weighting all n_c observations in class category c by the value $1 / n_c$ (e.g., Greenblatt, Mozdzierz, Murphy, & Trimakas, 1992; McLachlan, 1992; Meehl & Rosen, 1955; Rorer & Dawes, 1982; Widiger, 1983).

For an application with C ≥ 2 class categories, a mean PAC of (1 / C) × 100% is expected by chance under the null hypothesis, that the attribute is uniform random (e.g., Carmony, Yarnold, & Naeymi-Rad, 1998; Yarnold & Soltysik, 1991a). This enables the computation of a standardized index of effect strength—the *effect strength for sensitivity,* or *ESS*—that may be used to directly contrast different ODA models, regardless of possible *structural* (number of class categories, attribute metrics) and/or *configural* (relative class category sample size imbalances, total sample sizes) differences. First, compute C* = 100 / C. Then, ESS is computed via the following *interactive transformation* of the mean PAC (equations used later in the chapter are numbered on the right):

$$\text{ESS} = (\text{mean PAC} - C^*) / (100 - C^*) \times 100\%. \qquad [1]$$

On this scale, 0 represents the theoretical lower bound (the classification performance expected by chance), and 100 represents the theoretical upper bound (perfect classification). For the hypothetical example, the effect strength for sensitivity is (70 − 50) / (100 − 50) × 100%, or 40%. Thus, the model provides 40% of the theoretical possible improvement in classification accuracy that it is possible to attain after removing the effect of chance. To facilitate clarity, imagine that the mean PAC had instead been 75% in the hypothetical example—that is, the value lying midway between chance (50%) and ceiling (100%) performance. In that case, the effect strength for sensitivity would have been 50%, corresponding to a return of 50% of the possible theoretical improvement versus chance.

Note that the effect strength for sensitivity of a *degenerate* model—one that classifies all observations into a single class category—is always zero. For example, consider a two-category application in which all observations were classified as a 1 (or, if you wish, as a 0): The mean PAC across classes = (100 + 0) / 2, or 50%. Or, consider a three-category application classifying all observations as a 2 (or 0 or 1): The mean PAC across classes = (100 + 0 + 0) / 3, or 33.3%. In our experience, degenerate conventional models (in particular, logistic regression models) are not uncommon in applications involving a class category with a disproportionately small proportion of the total sample of observations. For example, in an application with 90 Class 0 observations and 10 Class 1 observations, an impressive overall PAC of 90% is obtained simply by classifying all observations as Class 0. This strategy yields perfect classification for Class 0 observations, but is perfectly useless for classifying Class 1 observations. Weighting by prior odds maximizes mean PAC rather than overall PAC, thereby reducing the likelihood of a degenerate model.

The third classification accuracy index used to evaluate an ODA model is *predictive value,* abbreviated as *PV,* which is the percentage of correct classifications into each class category. PV reflects the model's ability to make correct classifications of observations into class categories. It is an index of the *prognostic utility* of an ODA model, reflecting its proficiency in accurately classifying observations into different class categories (Dawes, 1962). The model

PV for Class Category 1 is the percentage of the classifications of observations into Class 1 that were correct, and the model PV for Class Category 0 is the percentage of the classifications of observations into Class 0 that were correct (in epidemiology these values are called the *positive and negative PV*, respectively). For multicategory applications, the PV for class category c is the percentage of model classifications into class category c that were correct. For the hypothetical example, PV for Class 0 = (20 / 30) × 100% = 66.7%, and PV for Class 1 = (15 / 20) × 100% = 75%. Thus, when the model predicted that an observation was a member of Class Category 1, it was accurate 75% of the time, and when it predicted that an observation was a member of Class Category 0, it was accurate 66.7% of the time. The mean PV across classes = (75% + 66.7%) / 2 = 70.8%. As was true for mean PAC, for a design with C ≥ 2 class categories, a mean PV across classes of (100 / C) × 100% is expected by chance: *effect strength for PV* is thus obtained using [1], with mean PV substituted for mean PAC. For the hypothetical example, effect strength for PV = (70.8 − 50) / (100 − 50) × 100%, or 41.6%. Thus, the model provides 41.6% of the theoretical possible improvement versus chance in the ability to accurately classify observations into class categories.

Ostrander, Weinfurt, Yarnold, and August (1998) note that, in contrast to sensitivity, PV is influenced by the base rate of the class category c in the population and by the false-positive rate (i.e., the probability that the model will classify an observation into class category c when the observation is *not* actually a member of c). Thus, the utility of an ODA model should be assessed for different base rates (Wainer, 1991). For a model to be *efficient*, it must provide a PV greater than the base rate for c in the population of interest (Meehl & Rosen, 1955). For example, if it is known that 53% of the population are in c, then a model should yield a PV for c of greater than 53%. Otherwise, the decision-maker is better off not using the model, and instead guessing that every observation will have a .53 probability of being a member of c (i.e., use the degenerate model: predicted class = c). For any given base rate (b), the model efficiency for class category c is defined as:

$$((\text{PV for class } c) \times (b)) / ((\text{PV for class } c) \times (b) + (\text{FP for class } c) \times (1 - b)), \qquad [2]$$

where FP, the false positive rate = 1 − PV for class c.

To facilitate clarity we illustrate an *efficiency analysis*—which is *not* implemented in the current introductory software system. We begin by computing model efficiency for the hypothetical example, for classifications into Class Category 1 (PV for Class Category 1 = .75; 1 − PV = .25; FP = .25) and b ranging between .1 and 1 in increments of .1 (by definition, when b = 0, model efficiency = 0). As seen, for the hypothetical example the ODA model efficiency exceeds all base rates except zero and one. Therefore, the ODA model clearly provides superior PV versus chance for classifications into Class Category 1.

b	$1 - b$	PV × b	FP × (1 − b)	Model Efficiency for $c = 1$	Efficiency − b
.1	.9	.075	.225	.075 / (.075 + .225) = .25	.15
.2	.8	.15	.2	.15 / (.15 + .2) = .4286	.2286
.3	.7	.225	.175	.225 / (.225 + .175) = .5625	.2625
.4	.6	.3	.15	.3 / (.3 + .15) = .6667	.2667
.5	.5	.375	.125	.375 / (.375 + .125) = .75	.25
.6	.4	.45	.1	.45 / (.45 + .1) = .8182	.2182
.7	.3	.525	.075	.525 / (.525 + .075) = .875	.175
.8	.2	.6	.05	.6 / (.6 + .05) = .9231	.1231
.9	.1	.675	.025	.675 / (.675 + .025) = .9643	.0643
1	0	.75	0	.75 / (.75 + 0) = 1	0

Note that, versus chance, the model does best for base rates between .3 and .5. Not illustrated here, efficiency analysis requires parallel analysis for classifications into all C class categories.

Finally, for the hypothetical example, the model is marginally more useful in a prognostic (effect strength for PV = 41.6%) versus a descriptive (effect strength for sensitivity = 40%) capacity. The *total effect strength,* defined as the mean of the effect strengths for sensitivity and PV, is a standardized omnibus measure of the utility of the ODA model versus chance. In the hypothetical example, total effect strength = (40% + 41.6%) / 2 = 40.8%. The model thus yields 40.8% of the theoretical possible improvement in overall (descriptive and prognostic) classification performance versus chance. In our applied research using ODA as an analytic tool, we focus strictly upon ESS as a classification performance measure because it does not vary as a function of base rate, whereas total and PV measures vary over base rate.

An interesting area for future research in ODA involves developing objective procedures for evaluating the ecological significance of all three (i.e., sensitivity, PV, and total) effect strength measures. Thompson, Yarnold, Williams, and Adams (1996) suggest an initial rule of thumb: A *weak* effect strength is defined as effect strength values < 25%, a *moderate* effect strength is defined as effect strength values between 25% and 50%, and a *relatively strong* effect strength is defined as effect strength values > 50%. We further suggest that a *strong* effect strength corresponds to effect strength values > 75%, and a *very strong* effect strength corresponds to effect strength values > 90%.

How to Obtain an ODA Model

To understand how ODA identifies an optimal model, imagine the following hypothetical application. A naval recruiting officer desires a fast, easy, and unobtrusive method to predict whether a person who walks into the recruiting office and inquires about the Navy will be successfully recruited. On the basis of previous experience, the officer believes that people who are recruited into the Navy are generally more fascinated by big ships than are those who are not recruited. To test this a priori hypothesis, the next eight people who walked into the recruiting office were greeted and asked to respond to the question: "On a scale from 1 (never) to 100 (every day), how often do you think about big ships?" After spending a standard hour-long recruiting session with each person, the officer noted whether the person was recruited (coded as a 1), or not recruited (coded as a 0). In this example, score on the 100-point scale is the attribute, and recruiting outcome (0 versus 1) is the class variable. For this example, ODA will determine how to use information concerning a person's answer to the question about frequency of thinking about big ships to predict with maximum overall PAC whether a person will be recruited. Imagine that the following results were obtained:

Person's ID	Score on Thinking About Big Ships Question	Outcome of the Recruiting Session
a	35	0
b	45	0
c	55	1
d	65	0
e	75	0
f	85	1
g	95	1
h	99	1

The first step in determining the ODA model for these hypothetical data involves reorganizing the data along a continuum, as follows:

```
        a         b         c         d         e         f        g  h
----- | -------- | -------- | -------- | -------- | -------- | -------- | -------- | ----- Score
     30        40         50         60         70         80         90        100
```

In this diagram, letters identify the eight people as well as the relative location of their score on the thinking about big ships question relative to the scores of all eight people. The next step involves replacing the letters, substituting the code "1" for people who were recruited, and the code "0" for people who were not recruited:

```
        0         0         1         0         0         1        1  1
----- | -------- | -------- | -------- | -------- | -------- | -------- | -------- | ----- Score
     30        40         50         60         70         80         90        100
```

We are now ready to identify the ODA model, which will be some specific combination of cutpoint and direction (below we illustrate the use of *brute force* to obtain the model—a practice that is becoming in vogue in applied research; e.g., Balogh & Merritt, 1996; Rothke et al., 1994).

A *cutpoint* is a point on the continuum that lies midway between successive observations that are from different classes. The three different cutpoints for the present example are indicated using arrows and identified using capital letters below [the value of Cutpoint A = (45 + 55) / 2 = 50; Cutpoint B = (55 + 65) / 2 = 60; and Cutpoint C = (75 + 85) / 2 = 80]:

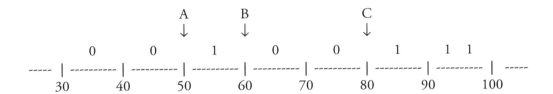

Direction refers to the manner in which cutpoints are used to classify observations, and is analogous to the "sign" of a correlation coefficient. There are two directions: *greater than* (observations with scores > cutpoint are predicted to be from Class 1, and observations with scores ≤ cutpoint are predicted to be from Class 0); and *less than* (observations with scores ≤ cutpoint are predicted to be from Class 1, and observations with scores > cutpoint are predicted to be from Class 0). Because the officer hypothesized that recruits (coded as 1) would be more interested in big ships, they should score at higher levels on the attribute than non-recruits (coded as 0). Therefore, it is necessary to consider only the "greater than" direction (if the hypothesis were instead nondirectional, then both directions would need to be searched). Thus, identifying the ODA model involves evaluating the classification performance achieved using all three cutpoints with the greater-than direction.

First, using Cutpoint A (50), all observations with scores > 50 are predicted to be from Class 1: This rule incorrectly classifies two observations actually from Class 0 (d and e) as being from Class 1, and correctly classifies all four observations actually from Class 1 as being from Class 1. Also, all observations with scores ≤ 50 are predicted to be from Class 0: This rule correctly classifies two observations actually from Class 0 (a and b) as being from Class 0, and there are no observations actually from Class 1 with scores ≤ 50. Thus, the model for Cutpoint A and the greater-than direction is: If score > 50 then predict that class = 1; otherwise predict that class = 0. For this sample, this model correctly classified 2 of the 4 observations actually from Class 0, and correctly classified all 4 observations from Class 1: Overall PAC = 6 / 8 × 100% = 75%.

Similarly, for Cutpoint B and the greater-than direction, the model is: If score > 60 then predict that class = 1; otherwise predict that class = 0. This model misclassifies one observation actually from Class 1 (c) as being from Class 0, and misclassifies two observations actually from Class 0 (d and e) as being from Class 1, for a total of three misclassifications: Overall PAC = 5 / 8 × 100% = 62.5%.

Finally, for Cutpoint C and the greater-than direction, the model is: If score > 80 then predict that class = 1; otherwise predict that class = 0. This model only misclassifies one observation actually from Class 1 (c) as being from Class 0: Overall PAC = 7 / 8 × 100% = 87.5%.

The ODA model is that combination of cutpoint and direction which yields the greatest value of overall PAC. Thus, the ODA model is: If score > 80 then predict that class = 1; otherwise predict that Class = 0. This model yields overall PAC = 87.5%. Because this model classifies all 4 observations from Class 0 correctly, sensitivity for Class 0 = 100%. Because 3 of the 4 observations from Class 1 were correctly classified, sensitivity for Class 1 = 75%. Thus, mean sensitivity across classes = (100% + 75%) / 2 = 87.5%, and effect strength for sensitivity = 75% (you may wish to verify that this also is the effect strength for PV, and therefore also is the total effect strength).

Note that the ODA model is consistent with the a priori hypothesis made by the recruiting officer: People who say that they frequently think about big ships (indicated by high scores on the attribute) make the best recruitment prospects. Indeed, all observations with scores greater than 80 in the present example were successfully recruited. Clearly, no other combination of cutpoint and the greater-than direction can achieve a greater value for overall PAC than was identified using ODA.

Selecting Among Multiple Optimal Models

ODA may identify more than one optimal model in applications with overall PAC < 100%. For example, imagine a hypothetical application in which the objective is to determine whether college basketball players who are drafted by the National Basketball Association (NBA) can be discriminated from non-drafted college players on the basis of their free-throw average (FTA) in their senior basketball season in college competition. Further imagine that the following data were obtained for six randomly selected college players, three of whom were drafted:

NBA Draft Status	FTA in Senior Year
Not Drafted	.567
Not Drafted	.621
	------ Optimal Model 1: Cutpoint = (.621+.623)/2 = .622

Drafted .623
Not Drafted .682
------ Optimal Model 2: Cutpoint = (.682+.718)/2 = .700
Drafted .718
Drafted .812

ODA was used to determine whether drafted versus non-drafted players (two-category class variable) can be discriminated on the basis of their senior season FTA (continuous attribute). As shown, two optimal models were identified—each of which misclassified one player. The first model is: If the player's FTA ≤ .622, then predict that the player is not drafted; otherwise, predict that the player is drafted. This model misclassified one player who was not drafted, and correctly classified all players who were drafted. The second model is: If the player's FTA ≤ .700, then predict that the player is not drafted; otherwise, predict that the player is drafted. This model misclassified one drafted player, and correctly classified all players who were not drafted.

Both ODA models suggest that NBA-drafted college players have a greater senior season FTA than non-drafted players. Suppose that a coach, scout, agent, player, or fan wanted to use this result to predict whether a college basketball player will be drafted by the NBA: Which of the two ODA models should be used? Because both models result in the same overall PAC, some additional selection criterion is needed. A variety of selection heuristics for selecting among multiple optimal models is available (Yarnold & Soltysik, 1991a).

For example, the *sample representativeness* heuristic retains the ODA model that results in a pattern of relative frequencies of predicted class memberships that is most similar—relative to the other ODA models—to the actually observed pattern of relative frequencies of class memberships for the sample. For example, imagine an application with observed relative frequencies of .2 and .8 for classes A and B, respectively. Further imagine that ODA identified two optimal models: Model 1 resulted in a predicted relative frequency of .2 for Class A and .8 for Class B, and Model 2 resulted in any different pattern of predicted relative frequency for Classes A and B. Model 1 would be selected by this heuristic because it resulted in a pattern of predicted class relative frequencies that was most similar to the actually observed pattern.

In the basketball example the observed relative frequency of the drafted and non-drafted classes was identical, with a value of 3 / 6 or .5. The sample representativeness heuristic would select the model that resulted in predicted class relative frequencies that were closest to this observed value of .5. Model 1 resulted in a predicted relative frequency of .33 (2 / 6) for non-drafted players and .67 (4 / 6) for drafted players. And Model 2 resulted in a predicted relative frequency of .67 (4 / 6) for non-drafted players and .33 (2 / 6) for drafted players. Because the predicted relative class frequencies produced by both models deviated equally from the observed relative frequency of .5 for each class, neither model would be preferred using this heuristic. In ODA software this heuristic is the default secondary selection criterion if none is specified.

The *balanced performance* heuristic retains the ODA model for which the absolute difference between the sensitivity for the two class categories is smallest: It is appropriate when it is equally important to classify observations from both class categories. For example, imagine an application in which two optimal models were identified: Model 1 achieved 60% PAC for Class A and 80% PAC for Class B, and Model 2 achieved 70% PAC for both A and B. Model 2 would be selected if the balanced performance heuristic were specified, because the absolute difference between the PAC for Classes A and B for Model 2 (0%) is less than the corresponding difference for Model 1 (20%). Because the absolute difference between the PAC for drafted and non-drafted players was 34% for both models in the basketball example, neither model would be preferred on the basis of this heuristic (i.e., both models are equally attractive by this heuristic).

The *category sensitivity* heuristic retains the ODA model that achieves the greatest sensitivity for a user-specified category (c) of the class variable: It is appropriate when it is especially important to classify observations from a particular class category. For example, consider an application with two optimal models: Model 1 achieved 50% sensitivity for Class A and 90% sensitivity for Class B, and Model 2 achieved 90% sensitivity for Class A and 50% sensitivity for Class B. If one were primarily interested in accurately predicting observations from Class B, Model 1 would be retained, and if one's primary interest was in Class A observations, Model 2 would be retained. In the basketball example, if one were primarily interested in maximizing PAC for drafted players then Model 1 would be selected because it best classified the drafted players. Were one primarily interested in maximizing PAC for non-drafted players, Model 2 would be selected because it best classified them.

The class category of greatest interest to the user is typically dummy-coded using a 1, and the class category of lesser interest using a 0. In epidemiology, the model PAC for Category 1 observations is called *sensitivity*, and the model PAC for Category 0 observations is called *specificity* (Friedman, 1987). The distinction between sensitivity and specificity is not necessarily trivial. For example, in medical research some attributes are especially useful as indicators of disease, whereas other attributes are especially useful as indicators of health (cf. Kraemer, 1992). An analogy exists between these ideas and the notions of *hit rate* (sensitivity) and *power* (specificity) from signal detection theory (e.g., Green & Swets, 1966; King et al., 1997; Loke, 1989; Swets, 1992).

The *maximum separation* heuristic, also known as the *distance* heuristic, retains the ODA model for which the absolute difference between the value of the attribute at the optimal cutpoint versus the nearest bordering observation is greatest (maximum), and is only useful in applications involving an ordered attribute. The motivation underlying this criterion is the selection of a model achieving the maximum absolute interclass separation. In the basketball example, for Model 1 the absolute difference between the value of the attribute at the cutpoint (.622) and the nearest bordering observation (.621 or .623) is .001. Because the corresponding absolute difference for Model 2 is larger (.018), Model 2 would be selected had this heuristic been specified.

The *maximum mean PAC* heuristic, also known as the *prior odds* heuristic, is appropriate for *imbalanced* or *nonorthogonal* applications involving class categories with different numbers of observations. It retains the ODA model with maximum mean PAC. An analogy exists between this heuristic and the use of prior odds, group-prior probabilities, or antecedent probabilities in conventional discriminant analysis (McLachlan, 1992). The motivation underlying this heuristic is the desire to avoid degenerate models and to achieve acceptable levels of classification accuracy for all class categories. ODA analyses involving weighting by prior odds explicitly maximize mean PAC: For such designs, this heuristic is superfluous. This heuristic is default for nonweighted applications.

For example, imagine an application with 10 Class A observations, 30 Class B observations, and two optimal models each resulting in 10 misclassifications (overall PAC = 75%). Model 1 misclassified 5 Class A observations (sensitivity for Class A = 50%) and 5 Class B observations (sensitivity for Class B = 83.3%): mean PAC = 66.7%. Model 2 misclassified 3 Class A observations (sensitivity = 70%) and 7 Class B observations (sensitivity = 76.6%): mean PAC = 73.3%. The maximum mean PAC heuristic would select Model 2 because it had a greater mean PAC than Model 1. In the basketball example, the mean PAC for both optimal models was 83.3%. Thus, neither model would be selected on the basis of this heuristic.

The *maximum PAC* heuristic is appropriate for applications involving weights, and retains the ODA model with maximum overall PAC. For example, consider a weighted application for which two optimal models emerged that each attained 75% weighted PAC. If Model 1 achieved an overall PAC of 80% and Model 2 achieved an overall PAC of 82%, then Model 2 would be selected by this heuristic. Because the basketball example did not involve weights, this heuristic is superfluous. This heuristic is default for weighted applications.

Finally, if the user does not wish to specify any particular a priori selection criterion, but rather desires that an optimal model is randomly selected in the event of multiple solutions, then the *random selection* heuristic will randomly select an optimal model.

ODA software allows one to specify a primary and a secondary selection heuristic before data are analyzed (cf. Pavur, 1992). The primary selection heuristic should reflect one's major interest or most important objective. If multiple optimal models are found, the primary selection heuristic is used to select among them. The secondary selection heuristic is of lesser priority, and will be executed if and only if multiple optimal models exist after first using the primary selection heuristic. If multiple optimal models meet both the primary and secondary selection heuristics, ODA software reports the first such model identified.

Assessing Model Stability

To what extent is the classification performance achieved by an ODA model—developed using a specific sample, known as the training sample—indicative or representative of the classification performance that would be achieved by subsequently using that model to make classifications for another independent sample of observations? ODA software uses leave-one-out and hold-out methods for assessing model performance stability (e.g., Thompson, 1994).

The so-called *one-sample jackknife*—a procedure that involves the random splitting of a data sample into two pieces—originated as a methodology for determining bias and standard error of the estimate for conventional statistical procedures such as t and r (e.g., see Efron & Gong, 1983). A related family of so-called *bootstrap* methods involves iterative resampling of jackknife half-samples and cumulating a frequency distribution of the statistic under investigation (e.g., Efron & Tibshirani, 1986; Hinkley, 1983; Stine, 1990). Bootstrap procedures have been recommended for use in validity generalization research because of their ability to generate accurate estimates of the standard error of correlation coefficients (Switzer, Paese, & Drasgow, 1992). Although research examining the utility of bootstrap estimation in clinical diagnosis is mixed (Grove, 1985), bootstrap estimates may overestimate positive predictive value (PPV) when the population PPV is low, and may overestimate the base rate in applications involving a variable that has a low base rate for one of the response levels. This may represent an important problem for bootstrap estimation, because consideration of the antecedent probability in statistical classification increases prognostic accuracy (e.g., Finn, 1982; Meehl & Rosen, 1955). It is therefore interesting that bootstrap methods have been proposed as a method for improving "cutting scores" for classifying observations when local base rates for one's sample differ from the base rates for the sample upon which the original cutting scores were developed (Rorer & Dawes, 1982).

The *leave-one-out* (abbreviated as *LOO*) procedure represents an extreme variation of the one-sample jackknife. The rationale for LOO validity analysis is that training classification performance reflects an optimistic, *upper-bound estimate* of the true cross-generalizability of the model. This is because of the problem of *overfitting*—that is, the training model capitalizes on chance errors that occur in the training sample to maximize classification performance specifically for the training sample. To the extent that independent samples do not share the idiosyncrasies of the training sample, the classification performance obtained when using the training model to classify observations in independent samples will generally be lower than was obtained in the training analysis.

LOO analysis reduces bias that occurs when estimating model performance using a single sample (Dunn & Vardy, 1966; Frank, Massy, & Morrison, 1965; Fukunaga & Kessell, 1971; Hills, 1966; Lachenbruch, 1967, 1975; Lachenbruch & Mickey, 1968). LOO analysis is widely used with time-series and discriminant analyses and gives an almost unbiased estimate of the confidence intervals for the true classification error rate (Hand, 1983; Hinkley, 1983). Other advantages of LOO validity analysis include that it is robust over violations of the normality assumption (Eisenbeis & Avery, 1972; Kshirsagar, 1972) and the relative ease with which it is generalized to multicategory applications (Eisenbeis, 1977).

In order to illustrate how LOO validity analysis proceeds, imagine an application involving ten observations. For identification purposes, index each observation using a unique integer between 1 and 10, inclusive. Thus, observations will be referred to as observation 1, observation 2, and so on. LOO is an iterative procedure, involving the same number of iterations as there are observations in the sample. Thus, in this example, the LOO procedure will require ten iterations.

In iteration number 1, (a) remove (i.e., hold out) observation 1 from the sample; (b) obtain an ODA model for the subsample consisting of observations 2–10; (c) use the resulting model to classify observation 1; and (d) store the result for later tabulation.

In iteration number 2, (a) hold out observation 2; (b) obtain an ODA model for the subsample consisting of observation 1 and observations 3–10; (c) use the resulting model to classify observation 2; and (d) store the result for later tabulation.

In iteration number i, (a) hold out observation i; (b) obtain an ODA model for the subsample consisting of all observations except for observation i; (c) use the resulting model to classify observation i; and (d) store the result for later tabulation.

Continue this procedure until every observation has been held out and classified on the basis of an ODA model that was obtained using a sample *that did not include the observation being classified*. In this example, the iterations would be completed after observation 10 was held out and then classified on the basis of an ODA model obtained for the subsample consisting of observations 1 through 9. Once iterating has terminated (a) tabulate the ten LOO classifications using a contingency table in which rows indicate observations' actual class membership, and columns indicate observations' predicted class membership; and then (b) compute the LOO classification performance of the model. If LOO classification performance is lower than training classification performance, this suggests that the ODA model has unstable classification performance that may decrease when the model is used to classify an independent random sample. In contrast, if LOO and training classification performance are the same, this suggests that the model has stable classification performance that may cross-generalize.

The most straightforward method for assessing the *classification error rate* of an ODA model is the *hold-out* cross-validation procedure, which is also referred to as *replication validity*. Estimating a model's hold-out validity simply involves (a) developing an ODA model for the training sample and (b) using that model to classify an independent hold-out sample of observations. The classification error rate observed in (b) is considered to reflect the true classification error rate for the model (e.g., Geisser, 1975; Stone, 1974). Although the estimated classification error rate achieved using the hold-out procedure is consistent and unbiased, this procedure requires relatively large samples as compared with alternative procedures, such as LOO or bootstrap (Eisenbeis, 1977). Because the hold-out procedure has no clear superiority to LOO, it is a less efficient error estimation procedure (Lachenbruch & Mickey, 1968). Note that, in LOO validity analysis, each observation constitutes a hold-out validity sample of size 1.

Standardizing Transformations

In contrast to parametric statistical methods (Bradley, 1968, 1978; Rasmussen & Dunlap, 1991), a particularly appealing characteristic of ODA is that the classification performance and associated Type I error obtained are invariant over any monotonic transformation of the attribute for a given data sample (Yarnold & Soltysik, 1991a). Nevertheless, sometimes context-specific standardization of attributes or weights is desirable or even necessary, particularly when conducting weighted, single-participant, or multiple-sample analysis.

In some applications it is appropriate to use the *absolute value* of a weight, rather than the raw value. For example, consider an application involving daily decisions to buy or sell a futures contract on the basis of some attribute: Were the contract to increase in value, a decision to buy the contract would have been profitable (and a decision to sell would have been unprofitable), and were the contract to decrease in value a decision to buy would have been unprofitable (and a decision to sell would have been profitable). Nonweighted ODA would maximize the percentage of the decisions (to buy or sell) that were profitable, but would not explicitly maximize the total return. To obtain maximum return, a weighted ODA should be used, with the weight being the absolute value of the price change in the contract. The absolute difference in the price of the contract is appropriate because, irrespective of what the decision (buy or sell) was, or whether that decision was profitable, it is the absolute value of the price change of the contract that reflects its underlying economic significance.

It is sometimes desirable to transform attributes or weights into a common scale, and an intuitive and frequently used technique for accomplishing this is *normative standardization* (Broverman, 1962; Cattell, 1952; Hicks, 1970; Saville & Willson, 1991; Yarnold, 1992). Under normative standardization, raw scores on the attribute or weight are transformed into corresponding standardized scores that reflect their magnitude relative to the population of scores—provided by a population of observations—for the attribute or weight. Thus, each observation is perceived as having scored at some percentile level relative to the population on the normatively standardized attribute or weight. To normatively standardize raw data use:

$$z_i = (x_i - \text{Mean}) / s_x ,$$ [3]

where z_i is the normatively standardized score for the ith observation, x_i is the raw value of the ith observation on the attribute or weight, Mean is the mean value of the attribute or weight for the sample, and s_x is the standard deviation of the attribute or weight for the sample. If, for example, the raw values on an attribute or weight were {1,2,3}, the corresponding normatively standardized scores would be {−1,0,1}, respectively. For any sample, the mean of the normatively standardized data will be 0, and the standard deviation will be 1.

An *ipsative standardization* is often useful when an observation has been scored on a single attribute or weight numerous times. Under ipsative standardization, raw scores on the attribute or weight are transformed into corresponding standard scores that reflect their magnitude relative to a population of scores—provided by the observation—on the attribute or weight (Broverman, 1962; Cattell, 1952; Clemens, 1966; Cunningham, Cunningham, & Green, 1977; Harris, 1953; Hicks, 1970; Jackson & Alwin, 1980; Mueser, Yarnold, & Foy, 1991; Rogers & Widiger, 1989; Saville & Sik, 1991; Saville & Willson, 1991; Yarnold, 1992). Thus, the observation is perceived as having scored at some percentile level on the attribute or weight relative to the population of scores for the observation on the attribute or weight. Raw data are ipsatively standardized using the formula for normative standardization, except that the Mean and standard deviation are computed based on the individual's scores, rather than on sample scores.

Finally, in an *interactive standardization* the attribute or weight is transformed into an *absolute scale* where, as an arbitrary example, 0 represents the theoretical minimum score (MIN) that can be attained and 1 represents the theoretical maximum score (MAX) that can be attained (Lamiell, 1981; Rogers & Widiger, 1989). For any raw score on the attribute or weight, denoted as X, interactive standardization is accomplished using the following formula:

$$X_{Interactive} = 1 - (MAX - X) / (MAX - MIN) , \qquad [4]$$

where $X_{Interactive}$ is the value of X on the 0 to 1 interactive scale (e.g., Yarnold, 1992). If, for example, an observation received a raw score of 2.5 on an attribute or weight for which the theoretical minimum score was 1 and the theoretical maximum score was 3, the corresponding interactive score would be 1 − (3 − 2.5) / (3 − 1), or .75.

CHAPTER 4

Evaluating Statistical Significance

*I*magine that you work on Wall Street, and your boss has assigned you the task of developing a classification model that would be used to guide investment decisions involving trading (buying and selling) stock on the New York Stock Exchange. Further imagine that the boss has tested your model by making the trades that it recommended for one month, that the model lost money, and that the boss has asked why your model performed so inadequately. How might the boss respond were you to explain that, "even though the model lost money, if the data do not violate the assumptions underlying the statistical procedure used to develop the model, then the performance level that the model achieved was nevertheless high enough to be considered highly unlikely if only chance was operating"? "That is," you further clarify, "even though my model lost money, it probably did better than would reasonably be expected if investment decisions were made randomly." Would the boss be satisfied by this explanation? Or, might the boss be more concerned with absolute aspects of your performance—that is, the bottom line? Would the situation be different if your task involved predicting whether depressed patients attempt to commit suicide, whether well-drillings produce oil, or whether a model successfully predicts X, where X = whatever? This scenario illustrates the difference between *practical* (also known as *clinical* or *ecological*) significance versus *statistical* significance.

Practitioners argue that the salient issue to them is practical significance—whether an outcome is sufficient to satisfy the patient, business, court, or whoever is evaluating the result. Practical significance is most logically conceptualized and evaluated in terms of the classification performance (overall percentage accuracy in classification [PAC], mean sensitivity, and PV) achieved using one's model in training and stability (LOO, hold-out) analysis. In the evaluation of practical significance, use of a weight may be intuitively appealing, as in the case of investment trading, where the profit made as a result of one's transactions is of greater interest than the percentage of one's decisions that were correct (profitable) versus incorrect (unprofitable). When the cost of misclassification is uniform across observations, primary interest may simply lie in maximizing overall effect strength.

Researchers argue that expert, intuitive, and visual ("eyeball") evaluation is subject to drift over time, can be unreliable within and between raters, and may be insensitive to small yet theoretically important statistically significant changes (e.g., Dawes, 1979; Feinstein, 1988; Goodman & Royall, 1988; Rosenthal, 1978; Simon, 1956; Yarnold, 1992). *Statistical significance* is most logically conceptualized and evaluated in terms of the Type I error rate (p) associated with the observed classification performance in training and validity analyses. Under the null hypothesis that the different class categories are not discriminable (e.g., the attribute is uniform random), p is the probability that a result as strong as or stronger than actually observed may have arisen by chance. For example, $p < 0.08$ indicates that classification performance as strong as observed occurs by chance less than eight percent of the time. Because p may attain any value > 0 (classification performance this strong will never happen by chance) and ≤ 1 (classification performance this strong will always happen by chance), it is necessary to establish a convention for determining when p is sufficiently small to consider the classification performance *statistically significant*.

It is conventional to consider $p < 0.05$ as indicating a significant effect. Although this convention is generally associated with Fisher (1925), its roots may be traced to the seventeenth century (Cowles & Davis, 1982). It has become commonplace for researchers to rely heavily on the presence or absence of statistical significance in evaluating the plausibility of a hypothesis, the success of an experiment, or the effectiveness of an intervention. However, current thinking favors the idea that overuse of p is misguided. First, violations by data of model assumptions—which for conventional statistical procedures are often inevitable—render the validity of p questionable (Bradley, 1968). Also, unlike effect strength, p does not speak to the issue of the *robustness* of an effect—only its *rarity*. Because of these shortcomings, it is argued that p be interpreted as an aid in summarizing evidence as opposed to establishing fact (Rosenthal & Rubin, 1985).

Such considerations notwithstanding, p remains an integral tool in contemporary scientific research. Accordingly, the development of a statistical framework for evaluating the classification performance achieved by an ODA model was an important step in popularizing the ODA paradigm (Carmony et al., 1998; Soltysik & Yarnold, 1994a; Yarnold & Soltysik, 1991a; Yarnold, Soltysik, & Martin, 1994). Two methods for obtaining p for an ODA analysis have been developed: the *analytic method* and *Fisher's randomization method*.

Analytic Methodology

The analytic method of determining p for an ODA analysis that we discovered is appropriate for nonweighted applications. In our own research (Yarnold & Soltysik, 1991a), we give an intuitive description of the analytic method and derive the theoretical distribution of optima for ODA applied to random data for a post hoc hypothesis and an ordered attribute. Our solution is open form (no formula) and has been enumerated for samples of up to 30 observations. We (Soltysik & Yarnold, 1994a) give an algorithmic description of the analytic method and describe the theoretical distribution of optima for ODA applied to random data for an a priori hypothesis and an ordered attribute: The solution is closed form and may be computed for huge samples. Carmony et al. (1998) present a mathematical proof of our distribution. Finally, we (Yarnold & Soltysik, 1992a) describe an open-form solution for the theoretical distribution of optima that arise when ODA is applied to random data, for a priori and post hoc hypotheses for any specific attribute measure metric.

To analytically derive the distribution of optima for post hoc ODA applied to an ordered random attribute, we begin by describing structural and configural features of the specific problem. For illustrative purposes, assume a binary class variable, a continuous random attribute, and three observations (two from Class Category 1, one from Class Category 0). The total number of observations is indicated as n, and the number of observations in class category c as n_c. We will now derive the theoretical distribution of optimal values for a post hoc, two-category, continuous random attribute ODA analysis, with $n_1 = 2$ and $n_0 = 1$.

First, it is necessary to determine the set of six possible outcomes that could occur if the attribute were, in fact, continuous and random (the two observations from Class Category 1 will be called "1A" and "1B"). The first possible chance outcome is: The value of the attribute for observation 1A is greater than that for observation 1B, which in turn is greater than that for the observation from Class 0. Symbolically, {1A > 1B > 0}. Similarly, the five remaining other possible chance outcomes include: {1A > 0 > 1B}, {1B > 1A > 0}, {1B > 0 > 1A}, {0 > 1A > 1B}, and {0 > 1B > 1A}. Because the attribute is random, each of these six possible outcomes is equally likely, with a probability of 1 / 6.

Because it is necessary to determine the optimal value for each of the six possible chance outcomes, ODA must be performed for each of the six data configurations. In the present example, two of the possible outcomes (those with observation 0 situated between observations 1A and 1B) have an optimal value of 2 correct classifications, and the other four possible outcomes have an optimal value of 3 correct classifications.

Finally, cumulating optimal values over the set of possible outcomes gives the theoretical distribution of optimal values for this specific ODA design. Here, the probability of achieving an optimal value of 3 is 4 / 6, and the probability of achieving an optimal value of 2 is 2 / 6. This is known as a *permutation probability* method, with permutations conducted over the attribute (cf. Berry & Mielke, 1992; Mielke, 1984; Westfall & Young, 1993). Yarnold and Soltysik (1991a) enumerated the theoretical distribution of optimal values for such problems with $n \leq 30$. For designs with $n > 28$, a Cray-2 supercomputer was required because of the enormous number of combinations to analyze.

A closed-form solution for the theoretical distribution of optima for a priori (versus post hoc) two-category ODA of continuous random data has been discovered, and exact p can be computed for such designs (Soltysik & Yarnold, 1994a). Examination of the a priori versus post hoc distributions reveals that, unlike many conventional statistical procedures, p for an a priori hypothesis is not simply one-half of the value of p for a *post hoc* hypothesis. For example, for balanced designs in which the classes have equal sample sizes, p (a priori) = $.5p$ (post hoc) only when overall PAC is 75% or greater. Although the analytic methodology generalizes to applications with more than two class categories, closed-form solutions have not yet been discovered for multicategory designs.

What if there are tied data, such that the continuity assumption is violated? As is true for conventional parametric statistics that assume continuity, such discontinuity in empirical data reflects the inherent human limitation of *imperfect measurement*, and thus does not compromise the theoretical probabilities (Bradley, 1968).

To investigate the distribution of optima for ODA applied to a binary random attribute, we proceeded with analytic enumeration. As was true for continuous data, for a binary attribute the solution for the post hoc distribution is open-form, the solution for the a priori distribution is closed-form, and it may be used to determine exact p for post hoc applications having sufficiently high overall PAC. Beyond these similarities, however, the continuous and binary distributions are different. This finding led us to the following two insights (Yarnold & Soltysik, 1992a).

First, there exists a theoretical dimension, which we refer to as the *precision dimension*, that may be used to describe the metric underlying the attribute for any specific application. The precision dimension is bounded at the extremes by binary data (least precise) and by continuous data (most precise). Just as exact distribution theory can be derived for the extreme cases on the precision dimension, so too can exact distribution theory be derived for any specific attribute measure metric. If, for example, the attribute were measured using a 7-point Likert scale, then the distribution theory may be derived assuming that a 7-point Likert scale was used. Because it is possible to derive distribution theory that assumes that the specific measure metric actually used in a given application was in fact used, it is clear that distribution theory for ODA can be based strictly on structural and configural features of a problem, and thus that such distribution theory will *never* be violated by data for a specific application (this is true for all permutation probabilities; cf. Mielke, 1984, 1991).

The second insight is that ODA clearly represents a powerful alternative to many of the most popular statistical tests. For example, Student's t test is traditionally used to analyze data consisting of a binary class variable and a continuous attribute. And chi-square analysis is widely used to analyze data consisting of a binary class variable and a binary attribute. Note that ODA can also be used, and exact distributions determined for, designs that lie *anywhere* on the precision dimension—that is, anywhere between the binary and continuous extremes. This is not true for conventional statistical procedures, however.

Fisher's Randomization Methodology

The permutation methodology used by the software for determining the exact p for ODA analyses is Fisher's randomization procedure, commonly used in generating permutation probabilities for a myriad of procedures (Bradley, 1968; Yarnold & Soltysik, 1992a). To illustrate this method, consider the following hypothetical data.

Actual Class

Observation ID	Category	Score on Attribute
A	0	10
B	0	20
		----- cutpoint = (20 + 30) / 2 = 25
C	1	30

There is a single optimal model: The cutpoint is 25, the direction is greater than (observations with attribute > 25 are predicted to be from Class Category 1; observations with attribute ≤ 25 are predicted to be from Class Category 0), and optimal value = 3 correct classifications. Using Fisher's method, one obtains every possible *shuffle* (permutation) of the class variable while holding the attribute constant:

Shuffle 1	Shuffle 2	Shuffle 3	Shuffle 4	Shuffle 5	Shuffle 6
A 0 10	A 0 10	B 0 10	B 0 10	C 1 10	C 1 10
---		---	---	---	---
B 0 20	C 1 20	A 0 20	C 1 20	A 0 20	B 0 20
---	---	---	---		
C 1 30	B 0 30	C 1 30	A 0 30	B 0 30	A 0 30

The first column under each shuffle is the observation's ID (all possible shuffles of observations are indicated); the second indicates the observation's class category membership status; and the third column indicates the attribute value (constant over shuffles). Next, ODA is performed for each shuffle: Four shuffles (1, 3, 5, and 6) have an optimal value of 3 correct classifications, and two shuffles (2 and 4) have an optimal value of 2 correct classifications. The dashes between class category membership codes indicate the predicted class memberships. In Shuffle 1, Class 0 observations are predicted to lie above the cutpoint (midway between 20 and 30, or 25), and Class 1 observations to lie below the cutpoint. Shuffle 2 illustrates two equivalently mediocre models. The top dashes suggest Class 0 observations lie above the cutpoint (15), and Class 1 observations lie below (observation B is misclassified). The bottom dashes suggest Class 0 observations lie below the cutpoint (25), and Class 1 observations lie above (observation A is misclassified). The distribution of optimal values is then cumulated over shuffles. Here, the probability of achieving an optimal value of 3 is 4 / 6, and the probability of achieving an optimal value of 2 is 2 / 6. For ODA analyses involving a single attribute, p by the analytic method is equivalent to p by Fisher's method.

Monte Carlo Methodology

Monte Carlo simulation is widely used in the study of statistical methods (Kalos, 1986; Kleijnen, 1974; Lehman, 1977; Mikhailor, 1992; Rubinstein, 1981). For example, conventional methods that are studied using simulation include chi-square (McClish, 1992; Schlundt & Donahoe, 1983) and other nonparametric procedures (Critchlow & Verducci, 1992), exploratory (Snook & Gorsuch, 1989) and confirmatory (Harris & Schaubroeck, 1990; Marsh, Balla, & McDonald, 1988) factor analysis, covariance structure analysis (Hu, Bentler, & Kano, 1992), univariate and multivariate analysis of variance (Stevens, 1992), multidimensional scaling (Isaac & Poor, 1974), bivariate (Switzer, Paese, & Drasgow, 1992) and canonical correlation (Barcikowski & Stevens, 1975), Markov models (Beck & Pauker, 1983), queuing networks (Rubinstein, 1986), multiple comparison (Hochberg & Tanhane, 1987; Seaman, Levin, & Serlin, 1991) and sequential testing (Mendoza, Markos, & Gonter, 1978) procedures, data transformations (Rasmussen & Dunlap, 1991; Saville & Willson, 1991), tests of normality (Wilk, Shapiro, & Chen, 1965), logistic regression (Anderson, 1972), and discriminant analysis (Markowski, 1990; Rubin, 1990a).

Monte Carlo simulation is not limited to the study of statistical procedures, however. Indeed, simulation has been used to study a vast array of issues in applications such as agriculture (Csaki, 1985), aircraft landing (Peters, 1985), archaeology (Hodder, 1978), chemical engineering (Ciccotti, Frenkel, & McDonald, 1987), cognitive abilities (Vale & Maurelli, 1983), courtroom backlog (Weiss, 1983), ecology (de Wit & Goudriaan, 1978), economic models (Pindyck & Rubinfeld, 1976), education (Crookall, 1987; Taylor, 1972), emergency systems (Sullivan & Newkirk, 1989), fertility (Santow, 1978), fire department operations (Carter, Chaiken, & Ignall, 1974), heart disease (Zhuo, Gatewood, & Ackerman, 1990), international relations (Gilboa, 1980; Guetzkow, 1963), logistics (Maggio, 1971), natural language (Vandamme, 1972), plant location planning (Churchill, 1969), operations research methods (Kohlas, 1972), particle physics (Binder, 1986, 1992; Lux, 1991), pedestrian movement (Turner, 1985), physiology (Modell, 1986), politics (Coplin, 1968), religion (Lilly, 1975), search, rescue, and emergency plans (Pritchett & Roehrig, 1985), simulation (Palarine, 1972), social systems (Cassidy, 1974), transportation (Gynnerstedt, Carlsson, & Westerlund, 1977; Hammesfahr, 1982), use of recreational facilities (Shechter & Lucas, 1978), and stressful work conditions (Tett et al., 1992).

Perhaps not surprisingly, Monte Carlo simulation is also useful in studying statistical aspects of ODA. Sometimes the number of shuffles of a class variable required by Fisher's randomization procedure to obtain a problem-specific exact permutation p is so large that it is computationally intractable. In such cases it is possible to use Monte Carlo simulation to estimate the p that would be obtained by performing Fisher's procedure. To perform such a simulation, first conduct ODA and obtain the *original optimal value.* Next, conduct Monte Carlo simulation involving a user-specified number of individual Monte Carlo experiments. For each experiment, the class category memberships of observations are randomly shuffled, ODA is conducted, and the Monte Carlo optimal value is compared with the original optimal value. The Monte Carlo p is the proportion of experiments in which the Monte Carlo optimal value equaled or exceeded the original optimal value.

Monte Carlo p is an estimate of the exact permutation p, and the accuracy of this estimate increases as a function of the number of Monte Carlo experiments conducted. Unfortunately, one cannot determine the accuracy of a given Monte Carlo p for some specific number of Monte Carlo experiments. However, one may determine the likelihood that an obtained Monte Carlo p is less than some specific "target" value. This likelihood is expressed in terms of confidence levels. For example, how confident is one that the exact p for a specific application is less than the target p (e.g., 0.05), given that the Monte Carlo p = 0.04 for 100 experiments? Clearly this confidence is less than would be the case had the Monte Carlo p = 0.04 for 1,000,000 experiments. Thus, the confidence that a Monte Carlo p is less than a target p increases with an increasing number of experiments. The method used to compute confidence levels by ODA software involves integrating the beta function by the method of partial fractions (cf. Noreen, 1989; Press, Flannery, Teukolsky, & Vetterling, 1989). Confidence levels reported by ODA software are rounded down to the nearest .01%, except in the case where 100% confidence is reported, in which case the computed value exceeds $1-e^{-28}$. We recommend that a confidence level of 99.9% or higher be used in actual applications.

Yarnold and Soltysik (1992a) conducted simulation research involving approximately two billion Monte Carlo experiments to investigate the accuracy, stability, and convergence properties of Monte Carlo estimation of exact p for known ODA distributions. Over the range of sample sizes (between 2 and 1,001 observations) and exact p (between 0.001 and 0.10) studied, 87% of the Monte Carlo p were identical to three significant digits to the corresponding exact p after 100,000 experiments (no estimates were inaccurate by more than 0.001). Also, 55% of the Monte Carlo estimates converged to their eventual value (i.e., for 100,000 experiments) after 7,000 experiments, and 86% of the estimates converged to their eventual value after 70,000 experiments. We typically use 10,000 Monte Carlo experiments in scientific publications (e.g., Grammer, Shaughnessy, & Yarnold, 1996; Harvey, Roth, Yarnold, Durham, & Green, 1996; Yarnold, Martin, Soltysik, & Nightingale, 1993).

Specifying the Type I Error Rate

Thus far we have been discussing how to go about determining or estimating p for a given application. We now focus attention on the issue of what *value* of p is appropriate for a given application. That is, what heuristic should be used to decide whether to accept or reject the null hypothesis—that class categories cannot be discriminated on the basis of the attribute—given the *empirical* (observed) p? Stated another way, what *target value* of p should one select as the criterion for determining whether the classification performance achieved by ODA in a given application exceeds the performance expected by chance?

The *generalized criterion* is appropriate for any study in which exactly one test of a statistical hypothesis is conducted. Discussed earlier, the conventional definition of both generalized ($p < 0.05$) and marginal ($0.05 < p \leq 0.10$) statistical significance is relatively arbitrary, and different criteria may be appropriate for different applications (cf. de Cani, 1984). For example, a relatively conservative criterion such as $p < 0.005$ may be appropriate in applications in which one has a great deal of power (e.g., large samples), or when committing a Type I error (i.e., concluding that an effect did occur when in fact that was false) would be very costly or otherwise undesirable. In contrast, a more liberal criterion such as $p < 0.15$ may be appropriate in applications in which one has relatively low power (e.g., small samples), or when committing a Type II error (concluding that no effect occurred when in fact that was false) would be very costly or otherwise undesirable.

Ryan (1959) distinguished between the *error rate per comparison* (the probability that any one of the comparisons—that is, tests of statistical hypotheses—conducted within a single study will be incorrectly considered to be significant) and the *error rate per experiment* (the expected number of such errors per experiment or study). The problem with using the generalized criterion to evaluate multiple statistical tests conducted within a single study is that the actual resulting criterion for the study (the *experimentwise p*) is higher (more liberal) than the generalized (per-comparison) criterion: an effect known as *alpha inflation*. Each time the significance of a statistical test is evaluated, the experimentwise p increases. For example, for c orthogonal contrasts the probability of committing at least one Type I error is equal to $1 - (1 - p)^c$ (Maxwell & Delaney, 1990). Thus, for a study involving eight orthogonal contrasts and a generalized (per-comparison) criterion of $p < 0.05$, experimentwise $p < 0.337$: For nonorthogonal contrasts, experimentwise p would be lower (see Stevens, 1992).

Many scientific articles report multiple tests of statistical hypotheses (i.e., *multiple comparisons*), yet evaluate the significance of each of the multiple statistical tests using only the generalized criterion. This situation is improving as researchers, reviewers, and editors become more concerned about this mismatch between theory and practice. However, use of the generalized criterion in applications involving multiple comparisons should not be completely abandoned. For example, imagine that a person is interested in only one of the multiple comparisons reported in a particular study. For that person, the generalized alpha criterion is appropriate with respect to that comparison because, from the perspective of that person, only that single test of a statistical hypothesis is actually evaluated. Such a scenario is often true in meta-analysis, which entails the quantitative review of literature in the context of specific research questions that may be addressed in a host of otherwise unrelated studies (Rosenthal, 1984). Accordingly, we recommend that effects meeting the experimentwise criterion for statistical significance (discussed later) are reported as "statistically significant," and that effects failing that but meeting the generalized criterion are reported as "nonsignificant trends" (e.g., Mueser et al., 1990, 1992).

There are two types of multiple comparison strategies: all possible comparisons versus planned comparisons. When conducting *all possible comparisons,* all of the different statistical hypotheses that it is possible to test (i.e., that can be constructed) for a given application are evaluated. In contrast, when conducting *planned comparisons* a theoretically circumscribed subset of all possible comparisons are evaluated. Whereas all possible comparisons are often conducted and reported in *exploratory* research, planned comparisons are *confirmatory*. No research yet addresses development of special-purpose methods for ensuring a specified experimentwise p in the context of the ODA paradigm. It is probably a sound bet that special-purpose methods will be developed for this purpose, analogously to the history of the development of special-purpose multiple comparison procedures for analysis of variance (e.g., Hochberg & Tamhane, 1987; Jaccard, Becker, & Wood, 1984; Keselman, Keselman, & Games, 1991).

Although ODA-specific special-purpose procedures are not yet available, it remains necessary to consider the number of statistical tests that are evaluated during the course of one's study and to control experimentwise p. Fortunately, a current multiple comparisons procedure may be adapted for the purpose of ensuring (or *protecting*) the experimentwise p in studies using the ODA paradigm.

Probably the most widely used and familiar of the general multiple comparisons methods is often referred to as the *Bonferroni procedure,* but this is a misnomer (e.g., Maxwell & Delaney, 1990). To make a long story short, Dunn (1961) applied an inequality derived by the Italian mathematician Bonferroni to the problem of ensuring a desired experimentwise p in multiple comparisons. She showed that the use of an adjusted per-comparison criterion—equal to the ratio of the desired experimentwise p (conventionally, $p < 0.05$) divided by the number of comparisons (nc) reported in the experiment—ensured the desired experimentwise p. Because Dunn's method is based on the Bonferroni inequality, her method is properly referred to as Dunn's Bonferroni-type multiple comparisons procedure (cf. Holland & Copenhaver, 1988).

Sidak (1967) showed that another Bonferroni-type inequality that holds independently of whether or not contrasts are orthogonal can be used to derive a per-comparison criterion of the form: $1 - \sqrt[nc]{(1 - p)}$, where p is the desired experimentwise p (conventionally, $p < 0.05$)

and *nc* is the number of contrasts (Maxwell & Delaney, 1990). As is true of Dunn's procedure, use of Sidak's per-comparison criterion ensures the desired experimentwise p. However, as seen in Appendix A, if the number of contrasts is greater than one, then the per-comparison p by Sidak's procedure will be greater (more liberal) than the corresponding per-comparison p by Dunn's procedure: a definite advantage when weak statistical power is a concern, and/or when the data are rare or expensive.

Sequentially rejective Bonferroni-type procedures have the benefit of ensuring a desired experimentwise p, yet they provide greater power than nonsequential alternatives (Holland & Copenhaver, 1987; Holm, 1979; Klockars, Hancock, & McAweeney, 1995; Shaffer, 1986). Although Dunn's and Sidak's procedures may both be adapted for sequential testing, as was true for nonsequential procedures, at any step of the sequential rejection procedure for which more than one outstanding test remains to be evaluated, the per-comparison p by Sidak's procedure will be greater than the corresponding p by Dunn's procedure. Because these sequentially rejective procedures are more easily illustrated than explained, the following example is presented in lieu of additional discussion.

As an illustration of the sequentially rejective Sidak Bonferroni-type multiple comparisons procedure (hereafter, the *sequential Sidak procedure*) for establishing statistical significance, imagine a study involving the evaluation via ODA of all three possible between-category comparisons for a design involving an ordered three-category class variable (lower, middle, and upper socioeconomic status) and an ordered six-category attribute (highest completion of grade school, high school, college, professional school, graduate school, or postgraduate school). Further imagine that experimentwise $p \leq 0.05$ is the desired criterion for establishing statistical significance, and that these results emerged:

Result of ODA Comparison	Corresponding $p <$
Education of upper-class > lower-class	0.01
Education of upper-class > middle-class	0.02
Education of middle-class > lower-class	0.04

To begin the sequential Sidak procedure, sort the observed ps in order of decreasing significance (increasing p): This has already been done above. The first step of the procedure involves evaluating the first observed p in the sorted list. To accomplish this, from Appendix A obtain the Sidak criterion for three tests (contrasts) and experimentwise $p < 0.05$. As seen, the Sidak per-comparison criterion is $p < 0.01696$. Because the first observed p (0.01) is less than the Sidak criterion (0.01696), the first p is considered statistically significant at an experimentwise $p < 0.05$. Had the first observed p exceeded the Sidak criterion, the procedure would terminate and it would be concluded that no comparisons were statistically significant. Because the first step did not terminate, the procedure continues.

To continue the procedure, we evaluate the second observed p in the sorted list. To accomplish this, from Appendix A obtain the Sidak criterion for two tests and experimentwise $p < 0.05$. As seen, the Sidak per-comparison criterion is $p < 0.02533$. Because the second observed p (0.02) is less than the Sidak criterion (0.02533), the second p is considered statistically significant at an experimentwise $p < 0.05$. Because the second step did not terminate, the procedure continues. Note that if, instead, the second observed p exceeded the Sidak criterion, the procedure would terminate, and it would be concluded that only the first comparison—between the upper and lower classes—was statistically significant at the experimentwise criterion. However, because the remaining tests are statistically significant at the generalized criterion (per-comparison $p < 0.05$), they would be considered nonsignificant trends. Also note that, had a nonsequential Sidak criterion been used, then the Sidak criterion for three tests (0.01696) would be used to evaluate all three comparisons—and the second observed p would *not* be considered statistically significant.

To continue the procedure, we evaluate the third (final) observed p in the sorted list. To accomplish this, use the Sidak criterion for one test and experimentwise $p < 0.05$. Because the third observed p (0.04) is less than the Sidak criterion (0.05), the third p is considered statistically significant at an experimentwise $p < 0.05$. Because no remaining observed ps remain to be evaluated, the procedure terminates. Note that, had the third observed p exceeded the Sidak criterion, the procedure would terminate and it would be concluded that only the first comparison (between upper- and lower-classes) and the second comparison (between upper- and middle-classes) were statistically significant.

A Priori Alpha Splitting

Until now, the various multiple comparisons under consideration were equally weighted: that is, the theoretical importance of each comparison was judged as being equivalent. This does not have to be true, however, and it is sometimes the case that some subset of all of the tests of statistical hypotheses evaluated in a given study are considerably more important to the central hypothesis than are the remaining tests. In such circumstances it is permissible to partition the experimentwise p in accordance with the perceived importance of the tests in the context of the study: a procedure referred to as *alpha splitting* (Maxwell & Delaney, 1990), or as *ensemble-adjusted* (Rosenthal, 1984; Ryan, 1985) or *ordered* (de Cani, 1984; Rosenthal & Rubin, 1984) *Bonferroni procedures*. It must be emphasized that alpha splitting is strictly an a priori methodology (de Cani, 1984; Maxwell & Delaney, 1990).

For example, imagine a two-phase study: Two confirmatory tests are considered in Phase 1, and nine exploratory tests are considered in Phase 2. Were the researcher to decide, on a strictly a priori basis, that the two confirmatory hypotheses were as important as the nine exploratory hypotheses, then each phase of the study would be assigned one-half of the desired experimentwise p. So, given a desired experimentwise $p < .05$, tests in Phase 1 would be evaluated *as though* experimentwise $p < 0.025$, and tests in Phase 2 would be evaluated *as though* experimentwise $p < 0.025$. At the conclusion of the analysis, overall experimentwise $p < (0.025 + 0.025)$, or $p < 0.05$, as desired.

Or, imagine a three-phase study in which Phase 1 hypotheses (involving 2 confirmatory tests) are judged, a priori, to be an order of magnitude (ten times) more important than the exploratory hypotheses evaluated in Phase 2 (10 tests) and Phase 3 (50 tests). Furthermore, because Phases 2 and 3 are judged to be of equal theoretical importance, they are assigned an equivalent overall portion of the experimentwise p (this is moot if Phases 2 and 3 each involve the same number of contrasts). In this example, if desired experimentwise $p < 0.05$, the p "reserved" for Phase 1 would be $(10 \times 0.05) / 12 = 0.04167$, and the p reserved for Phases 2 and 3 would each be $(1 \times 0.05) / 12 = 0.0042$. Here, the "ten" in parentheses for Phase 1 indicates that hypothesis in ten times as important as either Phase 2 or 3 ("one" each): hence, the sum (and thus denominator) is $10 + 1 + 1$, or 12. The overall experimentwise $p = (0.04167 + 0.0042 + 0.0042) = 0.05007$ (this differs from 0.05 due to round-off error).

Finally, consider an example of alpha splitting used in conjunction with a sequential Sidak procedure. Imagine you are investigating whether Type A behavior (a situation-specific response to stress involving attempts to assert control) is related to psychological instrumentality (a cognitive focus on getting the job done). Individuals (Type A and Type B) completed a ten-item survey measure of instrumentality, and responses to the ten items were summed to form a total score (I_t). Because none of the ten items was considered a perfect indicator of instrumentality, it was decided that the best test of the hypothesis that As and Bs can be discriminated on the basis of instrumentality involved a single ODA with I_t as the attribute. This single test constituted Phase 1 of the study. However, in order to determine whether specific components of instrumentality reliably discriminate As and Bs, it was decided to perform post hoc ODA on each of the ten items constituting the instrumentality questionnaire ($I_1, I_2, \ldots, I_{10}$). These ten tests constituted Phase 2 of the study. Before conducting the analyses it was decided to alpha split, with half of the desired experimentwise p assigned to each phase of the study. A conservative criterion was desired for each phase of the analysis ("phasewise" $p < .01$), such that experimentwise $p < .02$. A sequential Sidak procedure was used to establish statistical significance in Phase 2. Assume the following results obtained:

Attribute	ODA Result ($p <$)	Attribute	ODA Result ($p <$)
I_t	0.0021	I_6	0.0098
I_1	0.00092	I_7	0.073
I_2	0.19	I_8	0.0041
I_3	0.018	I_9	0.0084
I_4	0.00112	I_{10}	0.049
I_5	0.00114		

We begin with Phase 1 analysis, for which phasewise $p < 0.01$ (half of the experimentwise p). Because there is only one test in Phase 1, we simply compare the observed p (0.0021) against the phasewise alpha criterion (0.01). Because the observed p is less than the phasewise criterion, it is concluded that As and Bs could be discriminated using I_t scores, with experimentwise $p < 0.02$.

Switching to Phase 2 analysis, first sort the ten individual items in terms of increasing values for corresponding observed p:

Phase 2 Attribute	Sorted ODA Result ($p <$)
I_1	0.00092
I_4	0.00112
I_5	0.00114
I_8	0.0041
I_9	0.0084
I_6	0.0098
I_3	0.018
I_{10}	0.049
I_7	0.073
I_2	0.19

In Appendix A, the fourth column is appropriate for Sidak's procedure with experimentwise $p < 0.01$. This is fortunate, because 0.01 is the desired phasewise p.

In Step 1 of the procedure there are 10 outstanding contrasts, so the Sidak $p < 0.001005$. Because the observed p (0.00092) is lower than the Sidak p, the effect for I_1 is statistically significant with experimentwise $p < 0.02$. Similarly, the effects for I_4 (Sidak $p < 0.00112$) and I_5 (Sidak $p < 0.0013$) are statistically significant with experimentwise $p < 0.02$. However, because the observed p for I_8 (0.0041) is greater than the Sidak p (0.001435), the effect for I_8 and for all of the attributes that follow it in the sorted list are not statistically significant with experimentwise $p < 0.02$. Attributes I_8 through I_{10} in the sorted list are significant at the generalized criterion, and thus are nonsignificant trends.

Note that, had a nonsequential Sidak procedure been used to establish statistical significance in Phase 2 analysis, then only the effect for I_1 would be considered significant with experimentwise $p < 0.02$. Also, had a sequentially rejective Dunn's procedure been used to establish statistical significance in Phase 2 analysis, only the effect for I_1 would be significant with experimentwise $p < 0.02$. This would also be the conclusion had a nonsequential Dunn's procedure been used. In contrast, using the conventional generalized criterion ($p < 0.05$), only the effects for I_2 and I_7 would be considered statistically *non*significant. Finally, it is interesting to consider what the results would have been had alpha splitting not been used. Regardless of whether Dunn's or Sidak's procedure had been used, or whether the analysis was sequential or nonsequential, with a total of 11 tests, *none* of the effects would be considered statistically significant with experimentwise $p < 0.02$. This latter result illustrates how one additional test can render all of one's findings statistically nonsignificant.

CHAPTER 5

Two-Category Class Variables

*T*he simplest ODA design involves a class variable with two categories: a so-called two-category, binary, nominal, or dichotomous class variable. For example, one might be interested in discriminating between people who are happy versus sad, or between profitable versus unprofitable motion pictures. More generally, we may be interested in discriminating between two groups of observations who either (a) meet versus fail to meet some criterion, such as symptom present versus absent, or (b) meet either one or the other of two different criteria, such as mother versus father. This chapter discusses the use of ODA to analyze data in two-category applications.

Applications Involving Binary Attributes

Any random variable for which only two response alternatives exist—such as *yes* versus *no*, *true* versus *false*, *good* versus *bad*, *up* versus *down*, *X* versus *not X*, or *X* versus *Y*—is a binary attribute. Attributes may naturally be assessed using a binary scale—such as *pro* versus *con*, *National Football Conference* versus *American Football Conference*, or *male* versus *female*. Attributes may also be made binary artificially—such as *young* versus *old* (defined by an arbitrary age-based cutpoint), or *rich* versus *poor*. Binary attributes may be implicitly ordered by median- or mean-split procedures. For example, observations with scores greater than the median, mean, or other user-specified value are classified into one category, and observations with scores below this criterion are classified into another category. Binary attributes may also be implicitly qualitative: for example, preference for green versus orange.

It is well known that binary data are less precise than ordered data, and that nonparametric statistical procedures are less powerful than parametric procedures (cf. Maxwell & Delaney,

1993). Nevertheless, binary data may sometimes be more informative than ordered data (e.g., Larichev, Olson, Moshkovich, & Mechitov, 1995). For example, consider cholesterol level—typically an integer ranging between approximately 25 and 1500. Rather than considering specific integer values, physicians assess whether a patient's cholesterol is "high" (e.g., > 240): Aggressive treatment is appropriate if cholesterol is high. Thus, in applications where cholesterol is used as an attribute to predict physician decision-making, the binary variable (high versus not high cholesterol) may be more appropriate than the integer cholesterol level. In this case, physicians behave similarly to a thermostat, automatic transmission, or computer-based stock index arbitrage system—for which certain threshold cutpoints serve as triggers to initiate action. It does not seem unreasonable to predict that, as is true for a thermostat, behavior becomes unstable and less reliable as the relevant attribute approaches the threshold value. This perspective suggests the relevance of fuzzy set theory (Smithson, 1987), and initial explorations of the use of fuzzy logic in ODA appear promising (Rubin, 1992).

Irrespective of such considerations, data consisting of a binary class variable and a binary attribute are commonplace. Such data generally are analyzed using chi-square analysis (Pearson, 1900). An approximate statistic that evaluates the degree of statistical independence between the binary variables, chi-square should not be used when the expected value for any cell—formed by cross-tabulating the class variable and the attribute (an application with C columns and R rows has CR cells)—is less than 5 (e.g., Cochran, 1954; Mosteller, 1968; Yarnold, 1970).

Constructing a test problem in which ODA obtains perfect intergroup discriminability and for which chi-square is an invalid test statistic is simple. For example, imagine an application with six observations, in which (as hypothesized) all three Class A observations score at level 0 on the attribute, and all three Class B observations score at Level 1. Chi-square should not be used to analyze these data because the expected value is less than 5 for all four cells. Small samples are not the only gremlin for chi-square. For example, class sample-size imbalance limits power to detect statistical dependence and calls into question the validity of the estimated Type I error rate (Parshall & Kromrey, 1996; Reynolds, 1977). And an inherent limitation of chi-square is that directional (a priori) hypotheses are not possible. In contrast, none of these constitutes a problem for ODA (Yarnold & Soltysik, 1992a). For example, for the test problem the directional hypothesis that attribute values coded as 0 are predictive of Class Category A yields errorless classification and has exact (permutation) $p < 0.05$.

Example 5.1

As an example of a nondirectional (i.e., post hoc, exploratory, or "two-tailed") hypothesis involving a binary class variable and a binary attribute, consider Newmark's (1983) data on the gender of random samples of wasps and honeybees. The issue is whether the proportion of males and females is different for wasps versus honeybees. For ODA, the nondirectional alternative hypothesis is that gender (binary class variable) can be discriminated on the basis of type of bee (binary attribute), and the null hypothesis is that this is not true.

Gender	Type of Bee	
	Wasp	Honeybee
Female	40	103
Male	1,600	1748

Data were entered in free tabular format: Rows corresponded to gender (row 1 = female; row 2 = male), and columns to type of bee (column 1 = wasp; column 2 = honeybee). Because of imbalance in the number of male versus female bees, the ODA analysis involved weighting by prior odds—the default option. *Unless turned OFF by the user, every ODA analysis involves weighting by PRIORS.* ODA script required to perform the analysis is:

```
OPEN ex51.dat;
OUTPUT ex51.out;
FREE TABLE 2;
CLASS ROW;
CATEGORICAL ON;
LOO;
MCARLO ITER 10000;
GO;
```

The resulting ODA model was: If column = 1 (wasp), then row = 2 (male); otherwise, if column = 2 (honeybee), then row = 1 (female). Rewritten for clarity: If type = wasp then predict gender = male; otherwise, if type = honeybee then predict gender = female. Stable in LOO analysis, classification performance was weak in practical terms, but statistically significant because of the large sample (exact $p < 0.000003$). Thus, although a statistically greater proportion of honeybees (5.6%) versus wasps (2.4%) are female, the practical strength of this effect is only marginally greater than chance.

Classification Performance Index	Value (%)
Overall classification accuracy	48.78
Sensitivity (male)	47.79
Sensitivity (female)	72.03
Effect strength, sensitivity	19.82
Predictive value (male)	97.56
Predictive value (female)	5.56
Effect strength, predictive value	3.13
Total effect strength	11.47

Example 5.2

As an example of a directional (i.e., a priori, confirmatory, or "one-tailed") hypothesis involving a binary class variable and attribute, consider B. M. Yarnold's (1990) data on the voting behavior of 152 Republican and 245 Democratic members of the U.S. House of Representatives regarding the Refugee Act of 1980. Because the Act was sponsored by the Democrats, it was a partisan objective. Thus, the directional alternative hypothesis was that Democrats should vote in favor of the Act (pro), and Republicans should vote against it (con), and the null hypothesis is that this is not true.

Vote	Political Affiliation	
	Republican	Democrat
Con	118	78
Pro	34	177

Data were entered in free tabular format: Rows corresponded to the attribute (row 1 = con; row 2 = pro), and columns to the class variable (column 1 = Republican; column 2 = Democrat). Because of the different numbers of congresspersons in the two political parties, the analysis used weighting by prior odds. The a priori hypothesis implies that the data should fall in the major diagonal of the cross-classification table: that is, from the upper-left-hand corner to the lower right-hand corner. For ODA software, this hypothesis can be represented using the less-than direction (DIRECTIONAL < 1 2;) or the greater-than direction (DIRECTIONAL > 2 1;). Note that LOO analysis was not conducted because it is superfluous for applications involving a directional hypothesis and a binary attribute. ODA script used to conduct the analysis was:

```
OPEN ex52.dat;
OUTPUT ex52.OUT;
TABLE 2;
CATEGORICAL ON;
CLASS COL;
DIRECTIONAL < 1 2;
MCARLO ITER 10000;
GO;
```

As specified by the DIRECTIONAL command, the ODA model was, If row = 1 (con), then column = 1 (Republican); otherwise, if row = 2 (pro), then column = 2 (Democrat). Rewritten for clarity: If vote = con, then predict that political affiliation = Republican; otherwise, if vote =

pro, then predict that political affiliation = Democrat. As seen, classification performance for this model was moderate in practical terms, and statistically significant (exact $p < 0.000001$).

Classification Performance Index	Value (%)
Overall classification accuracy	72.48
Sensitivity (Republicans)	77.63
Sensitivity (Democrats)	69.41
Effect strength, sensitivity	47.04
Predictive value (Republicans)	60.20
Predictive value (Democrats)	83.89
Effect strength, predictive value	44.09
Total effect strength	45.57

Applications Involving Polychotomous Attributes

Polychotomous attributes such as race or religion are typically a challenge to analyze in conventional statistics (Bishop, Fienberg, & Holland, 1975). To simplify analysis employing a polychotomous attribute, researchers usually dissect it using a series of new, dummy-coded binary attributes. For example, imagine that race was measured using multiple categories: African American, Hispanic, Asian, and so forth. In conventional analysis, such as multiple regression or logistic regression, race would be analyzed using binary attributes: African American (coded as 1 if the observation is African American, and coded as 0 otherwise); Hispanic (coded as 1 if the observation is Hispanic, and coded as 0 otherwise); and so forth. In contrast, the analysis of polychotomous variables is straightforward in the ODA paradigm.

Example 5.3

As an example of a nondirectional hypothesis involving a binary class variable and a polychotomous attribute, consider Hawley and Wolfe's (1991) data on gender and rheumatic disease (RD), assessed using seven clinically defined, clearly different disorders. The question is whether men and women have different types of RD. For ODA, the nondirectional alternative hypothesis is that gender (binary class variable) can be discriminated on the basis of type of RD (polychotomous attribute): The null hypothesis is that this is not true.

Data were entered in free format (tabular input would have been much more efficient). Each of the 1,522 patients was represented using two space-delimited codes: gender (male = 1, female = 2) and then type of rheumatic disease (neck pain = 1, rheumatoid arthritis = 2, osteoarthritis of the knee = 3, low back pain = 4, osteoarthritis of the hands = 5, degenerative overlap syndromes = 6, fibromyalgia = 7). These disease categories are sorted by increasing percentage of female representation; however, this is *not* required by ODA software. The disease categories could have any random dummy index codes, and yet the same ODA model would be obtained. Data were weighted by prior odds. ODA script used to conduct the analysis was:

```
OPEN ex53.dat;
OUTPUT ex53.OUT;
VARS gender disease;
CLASS gender;
ATTRIBUTE disease;
CATEGORICAL disease;
LOO;
MCARLO ITER 10000;
GO;
```

The resulting ODA model was, If disease = 1, 2, 3, or 4, then gender = 1 (male); otherwise, if disease = 5, 6, or 7, then gender = 2 (female). Rewritten for clarity: If type of rheumatic disease = neck pain, rheumatoid arthritis, osteoarthritis of the knee, or low back pain, then predict that the person is a male; Otherwise, if disease = osteoarthritis of the hands, degenerative overlap syndromes, or fibromyalgia, then predict that the person is a female.

Classification Performance Index	Value (%)
Overall classification accuracy	46.12
Sensitivity (males)	77.34
Sensitivity (females)	37.45
Effect strength, sensitivity	14.79
Predictive value (males)	25.57
Predictive value (females)	85.60
Effect strength, predictive value	11.18
Total effect strength	12.98

Stable in LOO analysis, classification performance was weak in practical terms, but was statistically significant (Monte Carlo $p < 0.0001$). Thus, although a statistically greater proportion of women versus men have osteoarthritis of the hands, degenerative overlap syndromes, or fibromyalgia, the practical strength of this effect is only marginally greater than chance.

For Examples 5.1 and 5.2, ODA software reported exact p. However, exact p was not reported for this example because distributional formulations for multicategory ODA designs have not yet been discovered. Note from the output that—for 10,000 Monte Carlo experiments—estimated $p < 0.0001$, and there is 99.95% confidence that the desired (i.e., actual) $p < 0.001$. Imagine that you just finished an analysis and were mopping up loose ends. Also imagine that, for each of the total of ten ODA analyses that you will report in your study, you desire 99.9% confidence that experimentwise $p < 0.05$. Finally, imagine that the present example corresponded to the fifth-least-significant (i.e., greatest) p value in your study (from Appendix A, the Sidak criterion for experimentwise $p < 0.05$ and five remaining contrasts is $p < 0.01021$). To verify that the present analysis meets the sequentially rejective Sidak criterion for experimentwise $p < 0.05$, use MCARLO TARGET .05 SIDAK 5 STOP 99.9 ITER 10000.

Applications Involving Ordinal Attributes

Consisting of a relatively small number of graduated levels of the measured attribute, *ordinal scales* rank among the most broadly used type of measurement scale in all of science. Likert-type scales, typically involving between five and ten response options, may be most common (Nunnally, 1978). For example, one's socioeconomic status may be assessed using a three-point ordinal scale, with categories corresponding to lower, middle, and upper class. Also widely used, *ordinal categorical scales* consist of a relatively small number of qualitative categories that are ordered with respect to some theoretical factor (Kazdin, 1992). For example, at the end of a clinical trial, patients might be classified as being worse, unchanged, or better. Here, the three qualitative categories are worse, unchanged, and better; the theoretical factor is quality of clinical outcome; and the categories are ordered from lowest (worse) to highest (better) with respect to quality of clinical outcome.

What statistical analysis should one use to determine, for example, whether men can be discriminated from women on the basis of socioeconomic status, or whether two different therapies can be discriminated on the basis of clinical outcome? Because the metric underlying the attributes is ordinal, neither chi-square (nominal data) nor t test (interval or ratio data) is ideally appropriate (Clogg & Shihadeh, 1994). Conventional statistical methods used with such data include the Mann-Whitney U test (Grissom, 1994; Mann & Whitney, 1947) and the log-linear model (Gilbert, 1993; Hagenaars, 1990). However, excessive ties can complicate and compromise the U test (Cliff, 1995; Emerson & Moses, 1985; Gonzalez & Nelson,

1996), and maximum likelihood-based methods require large samples. As seen in the following examples, analysis of such data by ODA—for which neither the absence of ties nor the use of large samples is assumed—is straightforward.

Example 5.4

As an example of a nondirectional hypothesis involving a binary class variable and an ordinal (rank) attribute, consider Mendenhall and Reinmuth's (1974) data concerning the efficiency of two different auditing methods. To compare methods, 18 store accounts were audited—half via each method—and the number of accounting errors found in each store's account was recorded. Because number of errors is not normally distributed, stores were ranked between 1 (most errors found) and 18 (fewest errors found). The question is whether the two auditing methods found the same number of accounting errors. In ODA, the nondirectional alternative hypothesis is that the two accounting methods (two-category class variable) can be discriminated on the basis of the ranked number of errors each commits (ordinal attribute): The null hypothesis is that this is not true.

Data were entered in free format, with each row reflecting the data for a different store. Because this problem is balanced (class categories have equal size samples), weighting by prior odds would be superfluous. Thus, PRIORS is turned OFF in order to maximize overall classification accuracy. The ODA script for the analysis is

```
OPEN ex54.dat;
OUTPUT ex54.out;
FREE audit rank errors;
CLASS audit;
ATTRIBUTE rank;
PRIORS OFF;
MCARLO ITER 10000;
LOO;
GO;
```

The resulting ODA model was, If ranked number of errors ≤ 9.5, then predict that auditing method = 2; otherwise, predict that auditing method = 1. The objective of an audit is to identify accounting errors, and because audit method 2 was associated with identification of fewer errors, audit method 1 is thus the superior auditing method.

Classification Performance Index	Training Value (%)	LOO Value (%)
Overall classification accuracy	100.00	94.44
Sensitivity (audit method 1)	100.00	88.89
Sensitivity (audit method 2)	100.00	100.00
Effect strength, sensitivity	100.00	88.89
Predictive value (audit method 1)	100.00	100.00
Predictive value (audit method 2)	100.00	90.00
Effect strength, predictive value	100.00	90.00
Total effect strength	100.00	89.44

The model achieved perfect classification in training analysis (estimated $p < 0.0001$) but, as shown, the classification performance fell marginally in LOO validity analysis (exact $p < 0.00021$): One store audited using method 1 was misclassified, and overall effect strength fell from 100% to 89.4%. Nevertheless, the effect remained strong in practical terms.

Finally, note that ODA is used to evaluate Type I error for LOO validity results. For any LOO cross-classification table, the a priori hypothesis is that the observations will fall in the major diagonal (from upper-left to lower-right) of the table because we hypothesize that the ODA model will be stable in LOO analysis. Thus, assessing p for the LOO results presently involves using ODA to evaluate a two-category, binary attribute, priors-weighted, directional hypothesis: ODA software does this automatically for two-category analyses if LOO analysis is requested.

Example 5.5

As an example of a directional hypothesis involving a binary class variable and an ordinal attribute, consider Berry and Mielke's (1992) data on socioeconomic status (SES) and political affiliation. SES was assessed using two attributes, education and occupational prestige (each measured using *quintiles*), separately for each attribute for the total sample. The lowest fifth (bottom 20%) of the scores are coded using 1, the next-lowest fifth of the scores are coded using 2, and so forth until the final fifth (top 20%) of the scores are coded using 5. Here the directional alternative hypothesis is that Republicans will score at lower levels than Democrats (binary class variable) on both measures of SES (ordinal attributes), and the null hypothesis is that this is not true.

Data were entered in free format: First was political affiliation code (Democrat = 11, Republican = 34), followed by the education and prestige quintile scores, respectively. Because there were different numbers of Democrats and Republicans, weighting by prior odds was used: The resulting ODA model thus maximized mean sensitivity. The directional alternative hypothesis is that Republicans will score at lower levels than Democrats (binary class variable) on both measures of SES (ordinal attributes). The ODA script for this analysis is

```
OPEN ex55.dat;
OUTPUT ex55.out;
VARS politics educate prestige;
CLASS politics;
ATTRIBUTE educate prestige;
DIRECTIONAL < 34 11;
MCARLO TARGET .05 SIDAK 2 STOP 99.99 ITER 10000;
GO;
```

The DIRECTIONAL command specifies that Republicans will have lower SES than Democrats (this hypothesis may also be specified using the command: DIRECTIONAL > 11 34;). Because two tests of statistical hypotheses are being evaluated, the Sidak criterion of $p < 0.02532$ is used so as to ensure an experimentwise $p < 0.05$ (see Appendix A). If neither attribute has an ODA model with an associated p that is less than this Sidak criterion with 99.99% confidence, then neither test is statistically significant with experimentwise $p < 0.05$. Simulation terminates when the estimated p is less than the Sidak criterion with 99.99% certainty, or when 10,000 Monte Carlo experiments have been run, whichever happens first.

For education, the ODA model was, If educate ≤ 3.5 then politics = 34 (Republican); otherwise, if educate > 3.5 then politics = 11 (Democrat). For prestige, the ODA model was, If prestige ≤ 2.5 then politics = 34; otherwise, if prestige > 2.5, then politics = 11. Rewritten for clarity, if education is scored in the 60th percentile or lower, or occupational prestige in the 40th percentile or lower, then predict that the observation is a Republican; otherwise, predict that the observation is a Democrat. Evaluated first (because it had lowest p), the model for prestige met the Sidak criterion for experimentwise $p < 0.05$ with 99.99% certainty. Evaluated next (last), the model for education met the single-test Sidak criterion ($p < 0.05$) in training analysis. Both ODA models had stable classification performance in LOO analysis.

Classification Performance Index	Value (%)	
	Prestige	Educate
Overall classification accuracy	90.00	80.00
Sensitivity (Republican)	91.67	100.00
Sensitivity (Democrat)	87.50	50.00
Effect strength, sensitivity	79.17	50.00
Predictive value (Republican)	91.67	75.00
Predictive value (Democrat)	87.50	100.00
Effect strength, predictive value	79.17	75.00
Total effect strength	79.17	62.50

Considered together, these results suggest that, as was hypothesized, Republicans have strongly and statistically significantly lower SES than Democrats. However, the effect is stronger for occupational prestige than for education.

Example 5.6

As an example of a nondirectional hypothesis involving a binary class variable and an *ordinal categorical attribute*, consider Snyder, Wills, and Grady-Fletcher's (1991) data on outcome of marital therapy for unhappily married couples. The issue is whether different types of therapy are associated with different therapeutic outcomes. For ODA, the nondirectional alternative hypothesis is that therapy (binary class variable) can be discriminated on the basis of outcome (categorical ordinal attribute), and the null hypothesis is that this is not true.

Therapy Type	Divorced	No Change	Improved
Insight	3	22	4
Behavior	12	13	1

Data were entered in free format: First was therapy code (insight = 1, behavior = 2), followed by the outcome score (divorced = 1, no change = 2, improved = 3). Because there were different numbers of people in the two therapies, weighting by prior odds was used: The resulting ODA model thus maximized mean sensitivity. The ODA script for this analysis is

```
OPEN ex56.dat;
OUTPUT ex56.out;
VARS therapy outcome;
CLASS therapy;
ATTRIBUTE outcome;
MCARLO ITER 10000;
GO;
```

Rewritten for clarity, the ODA model was, If outcome = divorced, then predict behavior therapy; otherwise, if outcome = no change or improved, then predict insight therapy. Stable in LOO analysis, classification performance was relatively strong and statistically significant (estimated $p < 0.0049$). Thus, there was a relatively strong, statistically significant tendency for behavior therapy to be related to worse outcomes than insight therapy in this example.

Classification Performance Index	Value (%)
Overall classification accuracy	69.09
Sensitivity (insight therapy)	89.66
Sensitivity (behavior therapy)	46.15
Effect strength, sensitivity	35.81
Predictive value (insight therapy)	65.00
Predictive value (behavior therapy)	80.00
Effect strength, predictive value	45.00
Total effect strength	40.40

If analyzed from a traditional analytic perspective, we might naturally treat therapy as an independent variable, treat outcome as a dependent variable, and evaluate whether outcome scores varied as a function of therapy. The traditional role of the outcome and therapy variables was juxtaposed for ODA, but the model nevertheless clearly summarizes the relationship between these variables. ODA allows one to estimate the likelihood that information concerning the response distributions obtained by two independent randomly selected groups on the ordinal categorical scale enables reliable discrimination between groups, and it identifies a statistical model to accomplish the classification task optimally. If ODA reveals that the groups can be successfully discriminated, it may be inferred that the responses of the two groups are not representative of the same population.

Example 5.7

As another example of a nondirectional hypothesis involving a binary class variable and an ordinal categorical attribute, consider Holroyd, Nash, Pingel, Cordingley, and Jerome's (1991) data concerning outcomes of two headache therapies. As in Example 5.6, the issue is whether different types of therapy (binary class variable) are associated with different outcomes (categorical ordinal attribute). For ODA, the nondirectional alternative hypothesis is that therapy can be discriminated on the basis of outcome, and the null hypothesis is that this is not true.

Type of Therapy	Little or No Improvement	Moderate Improvement	Much Improvement
Cognitive–Behavioral	1	6	11
Amitriptyline	5	6	5

Data were entered in free format: First was therapy code (cognitive–behavioral = 1, amitriptyline = 2), followed by the outcome score (little or no improvement = 1, moderate improvement = 2, much improvement = 3). Because there were different numbers of people in the two therapies, weighting by prior odds was used: The resulting ODA model thus maximized mean sensitivity. The ODA script for this analysis is the same as for Example 5.6, except that "ex57.dat" is used in the OPEN command, and "ex57.out" is used in the OUTPUT command. Rewritten for clarity, the ODA model was, If outcome = much improvement, then predict cognitive–behavioral therapy; otherwise, predict amitriptyline. Although this suggests that cognitive–behavioral therapy was associated with more positive outcomes than was amitriptyline HCL therapy, the model was not statistically significant (estimated $p < 0.12$), and had only moderate practical significance.

Classification Performance Index	Value (%)
Overall classification accuracy	64.71
Sensitivity (cognitive–behavioral)	61.11
Sensitivity (amitriptyline)	68.75
Effect strength, sensitivity	29.86
Predictive value (cognitive–behavioral)	68.75
Predictive value (amitriptyline)	61.11
Effect strength, predictive value	29.86
Total effect strength	29.86

Classification performance was stable in LOO validity analysis for classifications involving cognitive–behavioral therapy, but it fell dramatically (to 31.2%) for classifications involving amitriptyline. This suggests that the model is highly unstable for classifications of amitriptyline, and that it is unlikely that classification performance will generalize if the model is used with independent samples. Interestingly, for the latter group, the attribute was uniformly distributed (5 patients reported little or no improvement, 6 reported moderate improvement, and 5 reported much improvement), reflecting the LOO sensitivity achieved for this group (31.2%). In contrast, the ratio of much-improvement to little-improvement is 11-to-1 for cognitive–behavioral therapy. Considered together, these findings suggest that, although the outcomes associated with amitriptyline therapy may be uniformly distributed (as expected by chance), the distribution of outcomes associated with cognitive–behavioral therapy suggests a possible therapeutic benefit. As indicated in the following example, use of a larger sample would likely have resulted in a statistically significant p value in the present example.

Example 5.8

As a final example of a directional hypothesis involving a binary class variable and an ordinal attribute, consider Hyde and Plant's (1995) data on the relative strength of gender effects versus other effects in psychology.

Type of Study	Ordinal Category of Cohen's d Statistic				
	0–0.10	0.11–.35	0.36–0.65	0.66–1.0	Over 1.0
Gender Differences	43	60	46	17	5
Other Effects	17	89	116	60	20

Note that, rather than use the continuous measure d, the range of d was *artificially parsed* into five unequal categories: The domain of the first (weakest effects) category is .10 d units, followed by .24 d units (second category), .29 d units (third category), .34 d units (fourth category); and the fifth category (strongest effects) has no upper bound for d. The question is whether published effects on gender differences are weaker than other published effects in psychology. The directional alternative hypothesis was that, compared to other effects, gender effects (binary class variable) should have weaker effect strengths (ordinal attribute).

Data were entered in free format: type of study (gender = 1, other = 2), then effect strength (0 to .10 = 1; .11 to .35 = 2; .36 to .65 = 3; .66 to 1.0 = 4; and over 1.0 = 5). Because of unequal numbers of studies, prior odds weighting was used: The resulting ODA model maximized mean sensitivity. The DIRECTIONAL command indicates that gender studies (coded as 1) should have lower effect strength scores than other studies (coded as 2). The ODA script for this analysis is

```
OPEN ex58.dat;
OUTPUT ex58.out;
VARS study strength;
CLASS study;
ATTRIBUTE strength;
DIRECTIONAL < 1 2;
MCARLO ITER 10000;
GO;
```

Rewritten for clarity, the ODA model was, If $d \leq .35$, then predict gender study; otherwise, if $d > .35$, then predict other study. ODA indicated that, relative to other areas, gender studies have disproportionately more effect sizes in the two lowest categories ($\leq .10$, and $.11 - .35$). As seen, the effect was weak in practical terms, and was statistically significant (estimated $p < 0.0001$).

Classification Performance Index	Value (%)
Overall classification accuracy	63.21
Sensitivity (gender study)	60.23
Sensitivity (other study)	64.90
Effect strength, sensitivity	25.13
Predictive value (gender study)	49.28
Predictive value (other study)	74.24
Effect strength, predictive value	23.52
Total effect strength	24.33

Finally, although effect strength was lower in this example (involving 473 observations) than the preceding example (involving 34 observations), p was lower (i.e., "more significant") presently. Whereas effect strength is standardized over sample size, p is not.

Applications Involving Continuous Attributes

One of the most frequently studied data configurations in all of science involves one two-category class variable and one continuous (interval or ratio) attribute. This design is ordinarily analyzed using between-groups t test, which evaluates the probability that an observed difference between the means on the attribute of samples representing the two classes could have arisen as a result of random error if the observations had been drawn from the same population (Student, 1908). It is easy to construct a test problem for which ODA suggests strong intergroup discriminability, and for which t test detects zero intergroup difference. For example, consider the following hypothetical data featuring one outlying Class A observation:

Observation ID	Class Membership	Attribute Value
1	A	−5
2	A	−4
3	A	−3
4	A	−2
5	A	29
6	B	1
7	B	2
8	B	3
9	B	4
10	B	5

Because the mean value of the attribute for classes A and B is identical (equaling 3.0), *t* test would conclude that the two classes could not be discriminated at all on the basis of the attribute. Note that even were *t* test to detect a significant difference between the means on the attribute of the two classes, no classification function (let alone optimal model) for predicting class membership would be given. In addition, the presence of a statistically significant mean difference as assessed using *t* test does not imply that the attribute will facilitate either ecologically or statistically significant classification accuracy. Outlying data are not the only gremlins for *t* test. Other structural features of applied data that are problematic for *t* test—serving to limit power to detect group mean differences, and calling into question the validity of the associated estimated Type I error rate (e.g., see Bradley, 1968)—include (a) class category sample-size imbalance (i.e., different numbers of 0s and 1s), (b) skewness in, (c) high intragroup variability in, and (d) non-normality of the attribute. None of these features of data constitutes a problem for ODA, however (cf. Yarnold & Soltysik, 1991a). For example, for the test problem, a directional ODA achieves 90% overall percentage accuracy in classification (PAC). The optimal model is, If attribute is less than 0, then classify the observation as a member of Class A; otherwise, classify the observation as a member of Class B (estimated $p < .039$).

Example 5.9

As an example of a directional hypothesis (for an example of a nondirectional hypothesis see DeArmon & Lacher, 1997) involving a binary class variable and a continuous attribute, consider Martin et al.'s (1987) data on heart rate variability (HRV) and susceptibility to sudden cardiac death (SCD). To determine whether people with depressed HRV are more susceptible to SCD, data were collected from 22 people who were susceptible to SCD and from 22 who were not susceptible. Because the class categories are balanced, PRIORS is set OFF, and an ODA model is sought that maximizes overall PAC. A continuous measure of heart rate variability—the Singer score—was computed for every person. Higher Singer scores indicate lower HRV. Thus, the directional alternative hypothesis is that, relative to people who are not susceptible, people who are susceptible to SCD (binary class variable) should have higher Singer scores (continuous attribute): The null hypothesis is that this is not true.

Data were entered in free format: first the class code (susceptible = 1; not susceptible = 0), then the continuous attribute (Singer score). The ODA script to conduct this analysis is

```
OPEN ex59.dat;
OUTPUT ex59.out;
VARS scd singer;
CLASS scd;
```

```
PRIORS OFF;
ATTRIBUTE singer;
DIRECTIONAL > 1 0;
MCARLO ITER 10000;
GO;
```

The resulting ODA model was, If Singer ≤ .0796, then predict that the person is not susceptible to SCD; otherwise, if Singer > .0796, then predict the person is susceptible to SCD.

Classification Performance Index	Value (%)
Overall classification accuracy	75.00
Sensitivity (not susceptible)	86.36
Sensitivity (susceptible)	63.64
Effect strength, sensitivity	50.00
Predictive value (not susceptible)	70.37
Predictive value (susceptible)	82.35
Effect strength, predictive value	52.72
Total effect strength	51.36

Classification performance, stable in LOO analysis, was relatively strong, and statistically significant (exact $p < 0.0037$). Consistent with the a priori hypothesis, higher Singer scores, indicative of lower HRV, are predictive of susceptibility to SCD.

Until now, we have discussed designs in which the cost of misclassifying an observation was identical—that is, unit-weighted—regardless of class category (nonweighted applications), or in which the cost of misclassifying an observation was identical within class category but different between category (applications involving weighting by prior odds). In applications involving an ordered attribute however, one may also specify *different misclassification weights* for individual observations. The weights may be personally subjective in order to obtain a *personalized* ODA model (cf. Dawes, 1979; Meehl, 1954). Thus, misclassifications of Class A observations might be stipulated to be twice (or any desired value) as important, desirable, or costly to you as are misclassifications of Class B observations (see the dating and vacationing examples, chap. 1, this volume). Alternatively, others may provide subjective weights. For example, in a market research application involving viewer ratings of the desirability of a television show, observations' estimates of the certainty of their responses could be used as weights: This would help ensure that data from people who were uncertain about their feelings toward the show did not influence results as strongly as did data from people who had strong feelings (see the beer brewing, personnel selection, and teaching examples, chap. 1, this volume). Finally, objective, application-specific weights may also be used. For example, when predicting whether it will rain, the most desirable ODA model might maximize total return in precipitation, and when predicting whether people will respond to direct market advertising, the most desirable ODA model might maximize the total dollar return of the mailing.

Example 5.10

As an example of a directional hypothesis involving a two-category class variable, a continuous attribute, and weighting by both prior odds and return, consider the *January Barometer*. According to Wall Street folklore, as the Dow Jones Industrial Index (DJII) goes in the month of January, so it goes for the entire year. For example, imagine that the DJII closed at 3,000 on the last trading day in December 1992. If the DJII closed *above* 3,000 on the last trading day in January 1993, then the January Barometer would be *positive*, and the DJII would be hypothesized to close above 3,000 on the last trading day in December 1993. In contrast, if the DJII closed *below* 3,000 on the last trading day in January 1993, then the January Barometer would be *negative*, and the DJII would be hypothesized to close below 3,000 on the last trading day in December 1993.

To evaluate this hypothesis, data were collected on closing values of the DJII for the last trading days in December and January between 1941 and 1991. First, we created a continuous January Barometer index (continuous attribute): JANCONT = (value of the DJII on last trading day of January, year Y) / (value of the DJII on last trading day in December, year $Y-1$) × 100. JANCONT values greater than 100 indicate that the DJII increased (and values less than 100 indicate it decreased) in value in January (not used, JANBIN is the binary indicator). Next, we created a two-category class variable (DECEMBER), that indicated whether the value of the DJII on the last trading day in December, year Y was greater (coded as 1) or lower (coded as 0) than the value of the DJII on the last trading day in January, year Y. For the fifty years represented in the data set, the DJII closed higher in 35 years and lower in 15 years. Because the problem is imbalanced, weighting by prior odds is indicated. Finally, the return weight (DJIIMOVE) indicates the percentage return that would have been achieved in a given year, had the correct decision—that is, to buy or sell the DJII—been made. The ODA script for this analysis is

```
OPEN ex510.out;
OUTPUT ex510.out;
VARS year janbin jancont december djiimove;
CLASS december;
ATTRIBUTE jancont;
WEIGHT djiimove;
DIRECTIONAL < 0 1;
MCARLO ITER 10000;
GO;
```

The resulting ODA model was, If JANCONT ≤ 100.69, then predict that DECEMBER = 0 (i.e., the market will decrease and you should sell); otherwise, if JANCONT > 100.69, then predict that DECEMBER = 1 (i.e., the market will increase and you should buy).

Classification Performance Index	Weighted Value (%)	
	Training	LOO
Overall classification accuracy	78.27	71.98
Sensitivity (market increase)	81.14	81.14
Sensitivity (market decrease)	70.70	47.79
Effect strength, sensitivity	51.85	28.94
Predictive value (market increase)	87.96	80.39
Predictive value (market decrease)	58.70	49.00
Effect strength, predictive value	46.66	29.39
Total effect strength	49.25	29.17

Classification performance in training analysis was relatively strong and statistically significant (estimated $p < 0.024$). Consistent with the a priori hypothesis, over the fifty years for which data were analyzed, the January Barometer significantly predicted priors- and return-weighted positive versus negative yearly DJII movement. However, because the performance achieved using the January Barometer fell to relatively weak levels in LOO validity analysis—particularly for years in which the market fell—the model may not generalize with comparable effect strength if used to classify data from subsequent years.

At this point we have completed discussion concerning problems involving dichotomous class variables, so let us briefly review what we did. Example 5.1 was simple, yet if the same data are analyzed using chi-square (for example), the solution obtained by ODA will not be obtained. Instead, a statistically insignificant effect—the opposite of what we report—would materialize because of the enormous skew in the data. Failure to adjust for the base rate is implicated in several well-known paradoxes (e.g., Simpson's, Lord's). Example 5.2 was straightforward, yet it accomplished something chi-square cannot do—test a one-tailed hypothesis in a problem involving binary variables. In Example 5.3, ODA found the best allocation of type of rheumatic disease to gender in one simple analysis. In contrast, the data were ill-conditioned for log-linear analysis because of numerous empirical zeroes, necessitating 21 separate bivariate analyses. Example 5.4 is straightforward by several statistical approaches, but ODA software provides a jackknife analysis to help assess overfitting. Examples 5.6 and 5.7 demonstrated how ODA can solve problems having too few observations and/or too many tied data values for analysis by traditional nonparametric methods. Example 5.8 illustrated an otherwise unavailable directional analysis for an ordinal measure. In Example 5.9, analysis via *t* test failed to show a between-groups effect as a result of skewed data, whereas ODA found a large difference with results invariant over any monotonic transformation of the data (not true for traditional methods). Finally, Example 5.10 illustrated a directional, double-weighted analysis that does not have a traditional analogue. It is already obvious that ODA can analyze data non-addressable by traditional analytic methods.

CHAPTER 6
Multicategory Class Variables

Designs involving class variables that have three or more levels or categories are known as multicategory designs (ODA software is capable of analyzing problems with an ordered attribute involving up to a maximum of ten class categories). For example, one might be interested in discriminating between different political parties, religious groups, or types of health insurance. In general, as the number of categories that one attempts to discriminate increases, the absolute level of classification error—for both humans and machines—also increases (e.g., Bacus & Gose, 1972; Sternberg, 1966). Nevertheless, as was true for two-category designs, for multicategory designs ODA finds a model that, when applied to a given sample of data, results in theoretically maximum (weighted or nonweighted) percentage accuracy in classification (PAC) for that sample. If desired, data may be weighted by prior odds and/or return, and hypotheses may be directional or nondirectional.

As in two-category problems, in multicategory problems ODA might discover multiple optimal models. The a priori selection heuristics for choosing among multiple optimal models for two-category problems also apply to multicategory problems. The category sensitivity, maximum mean PAC (prior odds), maximum PAC, and random heuristics generalize directly from the two-category case to the multicategory case. Whereas the sample representativeness, balanced performance, and maximum separation (distance) heuristics also generalize, we discuss them briefly because their functioning for the multicategory case is potentially ambiguous.

As in the two-category case, for multicategory problems the sample representativeness selection heuristic retains the ODA model that results in a pattern of relative frequencies of predicted class memberships that are most similar—relative to the other optimal ODA models—to the actually observed pattern of relative frequencies of class memberships for the sample. For each optimal model, for each class category, the absolute difference between the observed and predicted relative frequency is determined. Then, for each optimal model, the sum of these absolute differences is obtained across classes. The ODA model with the smallest (minimum) sum of absolute differences is retained by this heuristic.

Also as in the two-category case, for multicategory problems the balanced performance selection heuristic retains the ODA model for which the absolute difference between the sensitivities obtained for the C class categories is smallest (minimum). For a problem with C class categories, there are $C(C-1)/2$ unique pairs of different class categories. For example, for a problem with three class categories—A, B, and C, the $3(3-1)/2 = 3$ unique pairs of different class categories are AB, AC, and BC. For each optimal model, for each unique pair of different class categories, the absolute difference between their corresponding sensitivities is computed. Then, for each optimal model, the sum of these absolute differences is obtained across all pairs. The ODA model with the smallest sum of absolute differences is retained.

Finally, as in the two-category case, for multicategory problems the maximum separation (distance) selection heuristic retains the ODA model for which the minimum absolute distance between the value of the attribute at the cutpoint versus the nearest bordering observation is greatest (maximum), and is only useful in applications involving an ordered attribute. For a problem with C class categories, there are $C-1$ cutpoints. For each optimal model, for each cutpoint, the absolute distance from the cutpoint to the nearest observation is computed. The ODA model for which the minimum such absolute distance—across cutpoints—is greatest is retained by this heuristic.

Applications Involving Binary Attributes

Imagine that you wished to ascertain whether gender (male, female) and political affiliation (Democrat, Republican, or Independent) are related. Perhaps you were thinking of political class in the context of being a three-category class variable, and gender in the context of being a binary attribute? Things will probably work out better if you reverse your thinking, regarding gender as the class variable, and political affiliation as the polychotomous attribute. In general, *if the class variable has more levels than the attribute, reverse the roles of the class variable and the attribute.*

Applications Involving Polychotomous Attributes

Polychotomous data are widely reported in the sciences, and their analysis has been the focus of a great deal of research—perhaps the majority of which involves methods based on chi-square (Bishop, Fienberg, & Holland, 1975). It is easy to create a test data set for which such methods are inappropriate. For example, imagine a design with three class categories (A, B, and C), each with three observations. Further imagine that all three Category A observations scored a value of X on an ordered attribute, all three Category B observations scored a value of Y, and all three Category C observations scored a value of Z. Because of the small sample and sparse cross-classification table, methods based on chi-square are inappropriate. In contrast, these test data are ideal for ODA: For a nondirectional test of the alternative hypothesis that the three-category class variable can be predicted (discriminated) on the basis of the three-category attribute, effect strength = 100%, estimated $p < 0.01$ (this chapter will teach you how to verify this).

Perhaps the most frequently used traditional recourse for analyzing polychotomous data is the *log-linear* (or related) model. Although this is irrelevant to analysis via the ODA paradigm (i.e., do not worry if this makes little sense), the log-linear approach to the analysis of contingency (cross-classification) tables that are larger than the 2×2 tables created by crossing two binary measures requires the use of techniques such as recursive partitioning and iterative proportional fitting in order to identify combinations of square and/or rectangular subtables that meet the assumptions of quasi-independence (Bishop, Fienberg, & Holland, 1975; Goodman, 1968; Reynolds, 1977). In such analyses, the presence of both structural and empirical zeroes (i.e., cells in the table which by definition do not exist, versus cells which theoretically exist but into which no data fall, respectively) serves to complicate both the mechanics underlying the analysis as well as the interpretation of the findings (Hagenaars, 1990). Because these procedures involve the use of both likelihood-ratio and goodness-of-fit chi-square statistics, sparse tables involving relatively few data points are particularly troublesome (Hagenaars, 1990; Reynolds, 1977). Furthermore, imbalanced column or row marginals are also problematic, often necessitating the use of single- or double-centered, percentage- and/or angular-based standardization in order to employ methodologies such as ANOVA, Markov models, log-linear analysis, probit analyses, or latent structural modeling to analyze the structure underlying such tables (Bishop, Fienberg, & Holland, 1975; Hagenaars, 1990). The ODA approach is substantially more straightforward.

Example 6.1

As an example of a nondirectional hypothesis involving a multicategory class variable and a polychotomous attribute, consider data on congressional voting on the 1836 Pinckney Gag rule, which had historical implications in its effect on antislavery petitions. Bishop, Fienberg, and Holland (1975) note that these data are difficult to interpret because of the imbalanced marginals, and use these data to illustrate the use of row- and double-centered standardization in the analysis of contingency tables.

	Yea	Abstain	Nay
North	61	12	60
Border	17	6	1
South	39	22	7

For ODA the nondirectional alternative hypothesis is that vote (three-category class variable) can be discriminated on the basis of region of the country (three-category attribute): The null hypothesis is that this is not true. Data were entered in free tabular format: Columns indicated the congressman's vote (column 1 = yea; column 2 = abstain; column 3 = nay), and rows indicated the region of the country the congressman represented (row 1 = north; row 2 = border; row 3 = south). The ODA script needed to conduct this analysis is

```
OPEN ex61.dat;
OUTPUT ex61.OUT;
CATEGORICAL ON;
TABLE 3;
CLASS COL;
LOO;
MCARLO ITER 10000;
GO;
```

The resulting priors-weighted ODA model is, If row = 1, then column = 3; if row = 2, then column = 1; and if row = 3, then column = 2. Rewritten for clarity: If region = north, then predict that vote = nay; if region = border, then predict that vote = yea; and if region = south, then predict that vote = abstain.

Classification Performance Index	Value (%)
Overall classification accuracy	44.00
Sensitivity (yea)	14.53
Sensitivity (abstain)	55.00
Sensitivity (nay)	88.24
Effect strength, sensitivity	28.88
Predictive value (yea)	70.83
Predictive value (abstain)	32.35
Predictive value (nay)	45.11
Effect strength, predictive value	24.15
Total effect strength	26.52

As seen, the classification performance, stable in LOO analysis, was moderate and statistically significant (estimated $p < 0.0001$). Note that p is *not* provided for LOO in this example: ODA software only automatically produces p for LOO results involving two-category class variables. If p for LOO is desired, then treat the LOO contingency table as a categorical data set, and test the directional alternative hypothesis that the data fall in the major diagonal, using weighting by prior odds for imbalanced problems (e.g., emulate Example 5.2, but in the DIRECTIONAL command substitute "< 1 2 3").

Example 6.2

As an example of a directional hypothesis involving a multicategory class variable and a polychotomous attribute, consider Reynolds' (1977) data on the political affiliation status of a total of 1,852 high school students and their parents. The seven different political affiliations (and their dummy-codes) included strong Democrat (1), Democrat (2), Independent-Democrat (3), Independent (4), strong Republican (5), Republican (6), and Independent-Republican (7).

The Political Affiliation of Student	Parents 1	2	3	4	5	6	7
1	180	108	30	20	2	5	3
2	147	167	39	30	10	38	17
3	63	78	38	30	14	30	14
4	33	49	32	50	17	42	14
5	9	13	14	23	17	35	45
6	16	29	14	23	17	92	61
7	9	13	4	10	9	35	64

Data were entered using free tabular format: Rows reflected the dummy-coded score for the student's political affiliation, and columns reflected the corresponding dummy-coded score for parents' political affiliation. Because of class sample size imbalance, prior odds weighting is used. The directional alternative hypothesis is that family members tend to have comparable political affiliations, and thus the political affiliation of the student (the multicategory class variable) should be directly discriminable (predictable) on the basis of the political affiliation of the parents (the polychotomous attribute): The null hypothesis is that this is not true. Note that when the number of class categories is the same as the number of categories of the polychotomous attribute and a directional hypothesis is specified, LOO analysis is superfluous. Following is the ODA script required to perform this analysis.

```
OPEN ex62.dat;
OUTPUT ex62.out;
CATEGORICAL ON;
TABLE 7;
CLASS ROW;
DIRECTIONAL < 1 2 3 4 5 6 7;
MCARLO ITER 10000;
GO;
```

For this directional model, classification performance was weak in practical terms. That is, overall PAC = 32.8%; effect strength for sensitivity = 19.4%; effect strength for predictive value = 17.9%; and total effect strength = 18.6%. Nevertheless, this result was clearly statistically significant (estimated $p < 0.0001$) because of the relatively large sample size. Thus, there is a practically weak, statistically significant tendency for students to have the same political affiliation as their parents.

Applications Involving Ordinal Attributes

As for two-category problems, for multicategory problems any random variable having three or more response alternatives with corresponding numerical values that constitute at least an ordinal (rank) scale may serve as an ordered attribute in ODA.

Example 6.3

As an example of a directional hypothesis involving a multicategory class variable and an ordinal attribute, consider Bishop, Fienberg, and Holland's (1975) data on the visual acuity of human right and left eyes. A total of 7,477 women underwent eye tests. Separately for each eye, unaided distant visual acuity was assessed using an ordinal 4-point vision grading scale (1 = highest acuity, 4 = lowest acuity).

Right Eye Grade	Left Eye Grade			
	Highest (1)	Second (2)	Third (3)	Lowest (4)
Highest (1)	1,520	266	124	66
Second (2)	234	1,512	432	78
Third (3)	117	362	1,772	205
Lowest (4)	36	82	179	492

For ODA the directional alternative hypothesis is that a person's eyes tend to have comparable visual acuity. Thus, the acuity level of the right eye should be directly discriminable (predictable) on the basis of the acuity of the left eye: The null hypothesis is that this is not true. Note that a LOO analysis would be superfluous in this application. Data were entered using free format. For each observation there were two coded (1–4) variables: first the right eye acuity, and then the left eye acuity. Below are the ODA script used to conduct this analysis, and the classification performance summary.

```
OPEN ex63.dat;
OUTPUT ex63.out;
VARS righteye lefteye;
CLASS righteye;
ATTRIBUTE lefteye;
DIRECTIONAL < 1 2 3 4;
MCARLO ITER 10000;
GO;
```

Classification Performance Index	Value (%)
Overall classification accuracy	70.83
Sensitivity (grade 1)	76.92
Sensitivity (grade 2)	67.02
Sensitivity (grade 3)	72.15
Sensitivity (grade 4)	62.36
Effect strength, sensitivity	59.48
Predictive value (grade 1)	79.71
Predictive value (grade 2)	68.05
Predictive value (grade 3)	70.68
Predictive value (grade 4)	58.50
Effect strength, predictive value	58.98
Total effect strength	59.23

As seen, for this directional model the classification performance was relatively strong and statistically significant (estimated $p < 0.0001$).

Example 6.4

As another example of a directional hypothesis involving a multicategory class variable and an ordinal attribute, consider Thompson and Yarnold's (1995) data on perceived waiting time and patient satisfaction. In a phone interview, patients gave two ratings regarding their recent experience in the emergency department: whether the patient waited longer than, shorter than, or just as long as he/she expected to see the doctor, dummy-coded as 1, 3, and 2, respectively (time); and the patient's satisfaction with his/her experience in the emergency department, coded as 1 = poor, 2 = fair, 3 = good, and 4 = excellent (satis).

	Patient Satisfaction			
Perceived Waiting Time	Poor (1)	Fair (2)	Good (3)	Excellent (4)
Longer than expected (1)	58	97	179	94
Just as expected (2)	11	40	276	292
Shorter than expected (3)	5	16	133	373

The directional alternative hypothesis is that waiting time can be discriminated on the basis of satisfaction—the greater the patient satisfaction (ordinal attribute), the shorter the perceived waiting time (three-category class variable): The null hypothesis is that this is not true. Weighting by prior odds is used because of class category sample size imbalance. The ODA script used to conduct this analysis was

```
OPEN ex64.dat;
OUTPUT ex64.out;
VARS time satis;
CLASS time;
ATTRIBUTE satis;
DIRECTIONAL < 1 2 3;
MCARLO ITER 10000;
LOO;
GO;
```

For this directional hypothesis, the resulting ODA model was, If satis ≤ 2.5, then time = 1; if 2.5 < satis ≤ 3.5, then time = 2; if 3.5 < satis, then time = 3. Rewritten for clarity: If waiting time was shorter than expected, then predict that satisfaction rating = excellent; otherwise, if waiting time was just as long as expected, then predict that satisfaction rating = good; otherwise, if waiting time was longer than expected, then predict that satisfaction rating = fair or poor.

Classification Performance Index	Value (%)
Overall classification accuracy	51.08
Sensitivity (longer)	36.21
Sensitivity (just as long)	44.59
Sensitivity (shorter)	70.78
Effect strength, sensitivity	25.79
Predictive value (longer)	68.28
Predictive value (just as long)	46.94
Predictive value (shorter)	49.14
Effect strength, predictive value	32.18
Total effect strength	28.99

Classification performance for this model was of moderate strength, stable in LOO analysis, and statistically significant (estimated $p < 0.0001$).

Applications Involving Continuous Attributes

A frequently studied data configuration, referred to as a completely randomized design, involves one multicategory class variable and one continuous (interval or ratio) attribute (Winer, 1971). This design is ordinarily analyzed using one-way between-groups analysis of variance (one-way ANOVA), which, like t test, evaluates the probability that the observed differences between the means on the attribute of samples representing the multiple classes could have arisen as a result of random error if the observations had been drawn from the same population. It is easy to construct an example for which ODA suggests excellent intergroup discriminability, and for which one-way ANOVA detects zero intergroup difference. For example, consider the following test data set, which features outlying data, heterogeneity (Class B has zero variance), non-normality, and a small sample size (each data point represents a different observation's score on the attribute):

Class Category A	Class Category B	Class Category C
29	35	5
30	35	42
31	35	43
50	35	50

Because the mean value of the attribute for these three categories is identical (equaling 35), F = 0, and ANOVA would conclude that these categories could not be discriminated at all on the basis of the attribute. Note that, as was true for *t* test, even were ANOVA to detect a significant mean difference, no classification function (let alone optimal model) for predicting class membership is provided. Furthermore, detection of a statistically significant mean difference by ANOVA does not imply that the attribute will facilitate practically or statistically significant levels of classification performance.

In contrast, ODA achieves 83.3% overall PAC for this test problem (the optimal model is, If attribute < 33, classify the observation into Category A; if 33 < attribute < 38.5, classify into Category B; and if attribute > 38.5, classify into Category C), suggesting excellent intergroup discriminability on the basis of the attribute. As was true for *t* test, problematic characteristics of data for ANOVA—which limit its power to detect intergroup differences and call into question the validity of the associated estimated Type I error rate—include non-normality, skewness, kurtosis, and high intragroup variability in the attribute (Bradley, 1968). None of these features of data necessarily constitutes a problem for ODA (Yarnold & Soltysik, 1991a).

Example 6.5

As an example of a nondirectional hypothesis involving a multicategory class variable and a continuous attribute, consider Melaragno, Smith, Kormann-Bortolotto, and Neto's (1991) data on Alzheimer's disease (AD). Possible correlates of AD, which become impaired through aging, include sister chromatic exchange (SCE), an indicator of cellular response to DNA damage, and cell proliferation potential (CPP), an indicator of cell reproduction rate. Melaragano et al. sought to determine whether people with AD have different levels of SCE and CPP than people without AD. Data were collected for five patients with AD (Class Category 3), for five older adults without AD who were matched for age with the AD patients (Class Category 2), and for five younger adults without AD (Class Category 1). For each person, continuous measures of both SCE and CPP were obtained.

Data were entered in free format: The first variable is the three-category class variable (adstatus, coded using 1 through 3), followed by the continuous attributes SCE and CPP, respectively. Because class categories are balanced, PRIORS weights are turned OFF, and an ODA model is sought that maximizes overall PAC. The nondirectional alternative hypothesis is that the three class categories can be discriminated on the basis of SCE and/or CPP scores: The null hypothesis is that this is not true. In light of the small sample and corresponding weak statistical power, generalized $p < 0.05$ is used to establish statistical significance for every test of a statistical hypothesis considered in this example. Following is ODA script used for this analysis.

```
OPEN ex65.dat;
OUTPUT ex65.out;
VARS adstatus sce cpp;
CLASS adstatus;
ATTRIBUTE sce cpp;
MCARLO ITER 10000;
PRIORS OFF:
LOO;
GO;
```

Consider first the results for SCE. Although classification performance in training analysis was relatively strong (total effect strength = 64.3%), it was not statistically significant ($p < 0.11$), and LOO validity analysis indicated that the effect is highly unstable (total effect strength = 32.1%). Taken as a whole, these findings suggest that SCE data do not enable reliable discrimination of the three class categories.

Consider next the results for CPP. Training classification performance was relatively strong (total effect strength = 73.7%) and it was statistically significant (estimated $p < 0.025$). However, classification performance fell in LOO analysis (total effect strength = 43.8%).

Classification Performance Index	Value (%)	
	Training	LOO
Overall classification accuracy	80.00	60.00
Sensitivity (adstatus = 1)	100.00	80.00
Sensitivity (adstatus = 2)	100.00	100.00
Sensitivity (adstatus = 3)	40.00	0.00
Effect strength, sensitivity	70.00	40.00
Predictive value (adstatus = 1)	83.33	80.00
Predictive value (adstatus = 2)	71.43	50.00
Predictive value (adstatus = 3)	100.00	—
Effect strength, predictive value	77.38	47.50
Total effect strength	73.69	43.75

The resulting ODA model was, If CPP $\leq$ 160.5, then predict adstatus = 3; otherwise, if 160.5 < CPP $\leq$ 228, then predict adstatus = 2; otherwise, if CPP > 228, then predict adstatus = 1. Thus, CPP values were highest for younger adults without AD, lowest for older adults with AD, and intermediate for older adults without AD. However, the abysmal model performance for older adults with AD—particularly as occurred during LOO analysis, where no classifications into this category were made—suggests that this specific aspect of the omnibus ODA model is unreliable.

It is generally true, regardless of the metric underlying the attribute, that omnibus tests are too broad to result in a clear understanding of variable relationships or category differences. Of course, it is sometimes the case that all of the different class categories are predicted with high sensitivity: In this happy circumstance, the ODA model is unambiguously appropriate over the domain of the class variable. It is more common with applied data, however, for a multicategory ODA model to work well (high sensitivity) for several class categories, but not to do so well for other class categories. In this circumstance, follow-up analyses are necessary if one wishes to know the specific reasons why the omnibus test is statistically significant.

Accordingly, to determine precisely which aspects of the present omnibus model were statistically reliable, all possible pairwise comparisons were conducted. This was accomplished by appending the following ODA script at the end of the script given earlier for this example (note that the following analyses consider only the CPP data, and not the SCE data).

```
ATTRIBUTE cpp;
EX adstatus = 1;
TITLE comparing class categories 2 and 3;
GO;
EX adstatus = 2;
TITLE comparing class categories 1 and 3;
GO;
EX adstatus = 3;
TITLE comparing class categories 1 and 2;
GO;
```

For the first pairwise comparison contrasting categories 2 and 3, classification performance was moderately strong in training analysis (total effect strength = 51.2%), but it was not statistically significant (estimated $p < 0.87$), and in LOO analysis it degraded to a degenerate solution in which all patients were predicted to be from category 2 (total effect strength = 0%). Taken as a whole, these results suggest that older adults with AD are not reliably discriminated from older adults without AD on the basis of CPP data.

For the second pairwise comparison contrasting Categories 1 and 3, training classification performance was errorless and statistically significant (estimated $p < 0.0064$). Although classification performance degraded in LOO analysis, it remained relatively strong (total effect strength = 60%). The ODA model was, If CPP ≤ 230.5, then predict adstatus = 3; otherwise, if CPP > 230.5, then predict adstatus = 1. These results suggest that older adults with AD reliably have lower CPP values than younger adults without AD.

For the third and final pairwise comparison contrasting Categories 1 and 2, training classification performance was errorless and statistically significant (estimated $p < 0.0069$). Although classification performance degraded in LOO analysis, it remained strong (total effect strength = 81.7%). The ODA model was, If CPP ≤ 228, then predict adstatus = 2; otherwise, if CPP > 228, then predict adstatus = 1. These results suggest that older adults without AD reliably have lower CPP values than younger adults without AD.

In summary, there is a relatively strong and statistically significant tendency for younger adults without AD to have higher CPP values than older adults, regardless of their AD status. However, CPP does not discriminate older adults with AD from older adults without AD.

We have reached the end of the discussion concerning straightforward discrimination of multicategory class variables. In Example 6.1, ODA easily analyzed a problem that Bishop, Fienberg, and Holland (1975) were unable to complete using traditional means. In Example 6.2, ODA easily analyzed a problem with so many sparse cells that, for traditional methods, the validity of the estimated Type I error was unknown. Using ordinal attributes, Examples 6.3 and 6.4 illustrated confirmatory (one-tailed) tests lacking a traditional analogue. A hypothetical example was then presented for an application involving a continuous attribute for which ODA (but not t test, ANOVA, or Fisher's discriminant analysis) found an effect because, unlike traditional parametric methods, ODA is insensitive to outlying data. Finally, Example 6.5 demonstrated that ODA can identify effects in samples so small as to be considered marginal/insufficient for traditional analysis. Chapter 7 builds on techniques presented in this and the prior chapter and highlights the importance of the inherent ability of ODA to conduct confirmatory analysis of theoretically motivated hypotheses.

CHAPTER 7
Reliability Analysis

Broadly speaking, psychometrics—a vast discipline with a long and distinguished history—addresses the reliability (consistency) and validity (truth) characteristics of measured variables, regardless of the type of measurement devices or instruments that are used to collect data (e.g., Allen & Yen, 1979; Brown, 1983; Carmines & Zeller, 1979; Cromack, 1989; Ebel, 1979; Ghiselli, 1964; Gulliksen, 1950; Lord & Novick, 1968; Magnusson, 1967; Nunnally, 1978). The examples of ODA analogs to traditional psychometric analyses presented in this chapter are not a comprehensive accounting of the utility of the ODA paradigm in psychometrics. Clearly, there is a great deal more to be discovered. Nevertheless, this chapter reveals the ease and power with which the ODA paradigm may be adapted to numerous psychometric problems of great importance in the sciences, focusing on the concept of *measurement reliability*.

If a measurement methodology is perfectly reliable, then different observers using that methodology to quantify the magnitude of an attribute manifested by a single, unchanging object will obtain the identical score, irrespective of when or how often the measurements are taken. Similarly, a single observer recording multiple measurements of the unchanging object will always obtain the identical unchanging score. As a measurement methodology becomes increasingly unreliable, however, the scores obtained using that methodology to measure a single unchanging object begin to vary increasingly within and between observers, testings, and times.

Many different procedures for assessing the reliability of a measurement methodology have been derived, all of which provide an index known as the reliability coefficient (r_{tt}) that serves as a summary estimate of the stability, consistency, or precision of the measurement methodology. The theoretical maximum value that r_{tt} can attain is 1, corresponding to perfect reliability. The theoretical minimum value that r_{tt} can attain is 0, corresponding to the complete absence of reliability—or perfect unreliability, in which case measurements behave as though they reflected a random variable. Although empirically it is possible that r_{tt} can attain a negative value (so-called *antireliability*), because this is considered theoretically impossible, a negative r_{tt} is interpreted as indicating that the measurement methodology is unreliable. Thus, r_{tt} is assumed to equal 0.

In addition to providing a reliability coefficient theoretically bounded between 1 and 0, many of the various approaches to estimating the reliability of a measurement methodology share a common theoretical basis deriving from classical test theory (e.g., see Cronbach, Gleser, Nanda, & Rajaratnam, 1972; Gulliksen, 1950; Kuder & Richardson, 1937). Classical test theory assumes that any given observed score is constituted by (a) the true value of the attribute in the object of measurement (i.e., *true score*) and (b) unmeasured sources of variability (i.e., *error score*). That is, *observed score* = true score + error score. It is further assumed that, if a distribution of observed scores was obtained for a single object of measurement, the resulting distribution of error scores would have a mean of zero, would be normally distributed, and would be independent of the magnitude of the observed scores. For a randomly selected sample of observations, the reliability of a measurement methodology is generally defined as the ratio of the variance of the true scores for those observations divided by the total variance—that is, the sum of the variance of the true scores plus the variance of the error scores.

Inter-Rater Reliability

Numerous experimental procedures are used in the empirical assessment of reliability, and one of the most widely and frequently used of these procedures is the study of inter-rater reliability (e.g., Bishop et al., 1975; Fleiss, 1986; Reynolds, 1977; Saal, Downey, & Lahey, 1980; Woolson, 1987). Inter-rater reliability provides an index of the level of agreement between two or more observers (raters) who independently evaluate (rate) a sample of observations with respect to a single attribute. For applications involving multiple attributes, it is appropriate to compute a multivariate index of inter-rater reliability (Conger & Lipshitz, 1973; Novick & Lewis, 1967; Rae, 1991; Yarnold, 1984), though this is rarely done in applied research.

As an illustration of an application for which inter-rater reliability is important, consider Woolson's (1987) example in which two cardiologists—X and Y—independently evaluated the same set of 200 electrocardiograms. Each cardiologist classified each electrocardiogram as being normal, possibly abnormal, or definitely abnormal, and the following hypothetical data emerged:

	Cardiologist Y		
Cardiologist X	Normal	Possibly Abnormal	Definitely Abnormal
Normal	90	30	0
Possibly Abnormal	0	20	20
Definitely Abnormal	10	10	20

Do these data support the directional alternative hypothesis that the cardiologists agreed in their interpretations of the electrocardiograms? (The null hypothesis is that they did not agree.) To understand these data in the context of the alternative hypothesis, it is easiest to begin by first determining whether the cardiologists were in complete agreement—that is, if there is evidence of perfect inter-rater reliability. Had the cardiologists always agreed about the status of the electrocardiogram, then all of the events (joint ratings) represented in the cross-classification table would fall along the three cells that constitute the major diagonal. That is, if there were perfect inter-rater reliability, all of the electrocardiograms would be jointly classified by both cardiologists as being normal, as being possibly abnormal, or as being definitely abnormal: All other cells in the table would be empty—that is, would constitute empirical zeroes.

Examination of this table reveals that the two cardiologists agreed most frequently (90 events) when the electrocardiogram was classified as being normal. Most frequent disagreement occurred when the electrocardiogram was classified by either cardiologist as being possibly abnormal (60 events). Because not all of the data in this table fall within the major diagonal, the inter-rater reliability cannot be perfect. However, because some of the data do fall along the diagonal, there is some level of inter-rater agreement. How can these data be summarized to obtain a single estimate of inter-rater agreement?

Perhaps the most intuitive and frequently reported index of inter-rater reliability is the overall percentage agreement between raters—or overall percentage accuracy in classification (PAC; Bacus & Gose, 1972; Faraone & Hurtig, 1985; Nishikawa et al., 1983; Yarnold & Mueser, 1989). As discussed by Woolson, however, the percentage of inter-rater agreement is generally considered to be an insufficient index because—before the discovery of the ODA paradigm—there has been nothing against which to compare the observed level of inter-rater agreement:

> One such measure is simply the percentage of agreement.... (For the data in the above table), this statistic is simply 65% [100% × (90 + 20 + 20) / 200]. The problem with using this statistic is its interpretation. In particular, to what do we compare the 65%? About the only situation in which percentage of agreement is a useful and informative number is when each cardiologist had rated the entire sample into one and the same diagnostic group ... and the only descriptive figure is the percentage of agreement. (pp. 253–254)

Accordingly, the suitability of several traditional approaches for assessing the level of inter-rater agreement has been investigated. Chi-square and methods based on chi-square are not acceptable measures of inter-rater agreement because they measure all forms of association, including gross disagreement (Woolson, 1987); are sensitive to imbalance in the row marginals which, when severe, constrains the theoretical maximum attainable value of the reliability coefficient to less than one (Davenport & El-Sanhurry, 1991; Reynolds, 1977); and should not be used when the expected value for any cell in the table is less than five (Yarnold, 1970).

In an attempt to correct for these difficulties, Cohen (1960) introduced the kappa coefficient for nominal data (see also Posner, Sampson, Caplan, Ward, & Cheney, 1990). However, like chi-square-based indices, kappa—for which the test statistic is the same as that obtained by applying a chi-square statistic to a 2 × 2 table (Feingold, 1992)—only attains its theoretical upper limit when the marginal distributions for the raters are equal (Reynolds, 1977). Although it is possible to compute p for kappa (Berry & Mielke, 1985; Cliff & Charlin, 1991; Critchlow & Verducci, 1992; Fleiss, 1986; Woolson, 1987), in practice this often is not done, perhaps because this is not widely available in popular statistics software packages. As currently is true for the ODA paradigm, the criterion for evaluating the practical significance of kappa is based on arbitrary suggestions (Landis & Koch, 1977). Weighted kappa—computed for ordered ratings—is problematic because weights are completely arbitrary, and generalized kappa—used to assess agreement for more than two raters (the result is interpreted as an intraclass correlation coefficient of reliability: see Fleiss & Cohen, 1973)—is problematic when few raters are available (Davies & Fleiss, 1982; Fleiss, 1986; Lefevre et al., 1992; Lefevre, Feinglass, Yarnold, Martin, & Webster, 1993; Posner et al., 1990; Soeken & Prescott, 1986).

No doubt attributable at least in part to these difficulties, Pearson's (1900) correlation coefficient r remains one of the most frequently used statistical methodologies for estimating inter-rater reliability (cf. Royeen, 1989). This is true even for applications where the data clearly are *not* continuous—for example, applications where data consist of Likert-type ratings, which involve only a few (e.g., ten or fewer) possible response alternatives (cf. Russell & Bobko, 1992; Yarnold & Mueser, 1989; Zegers, 1991) and thus for which indices of ordinal association such as Kendall's (1938) tau coefficient or Spearman's (1904, 1910) rank correlation coefficient are appropriate (e.g., Woolson, 1987). The use of Pearson's r in this context is problematic because r assumes continuity as well as independent groups (the latter assumption is particularly troublesome in applications involving the estimation of temporal reliability).

Finally, an increasing trend in the estimation of reliability involves use of the intraclass correlation coefficient of reliability, an analysis of variance approach reflecting early classical test theory and modern generalizability theory traditions (Brennan, 1983; Cronbach, Glaser, Nanda, & Rajaratnam, 1972; Fisher, 1921; Fleiss, 1986). The intraclass correlation approach to estimating reliability is sensitive to violations of assumptions underlying its distribution theory (e.g., normality, homogeneity, and balance in row and column marginals), may require prohibitive amounts of computer memory for large problems, and may have low power when there is relatively little between-rater variation (e.g., Guyatt, Walter, & Norman, 1987; Hsu, 1990).

Fortunately, the ODA paradigm represents an intuitive and powerful alternative perspective concerning the assessment of inter-rater reliability. In the ODA approach, perfect inter-rater reliability is achieved when all of the data fall in the major diagonal of the table created by cross-classifying the ratings of the two observers. For ODA, evaluating this directional alternative hypothesis simply involves determining the percentage of (weighted or non-weighted) inter-rater agreement and evaluating this level of agreement for both statistical and practical significance. The null hypothesis is that there is no evidence of inter-rater agreement.

Example 7.1

As an example of inter-rater reliability analysis involving an ordinal attribute, consider Woolson's (1987) hypothetical data on two cardiologists, both of whom independently classified each of the same set of 200 electrocardiograms into one of three mutually exclusive and exhaustive diagnoses: normal, possibly abnormal, and definitely abnormal. The directional hypothesis is that, because cardiologist X's ratings and cardiologist Y's ratings are consistent—that is, fall into the main diagonal of the cross-classification table—cardiologist X's ratings (three-category class variable) are therefore directly discriminable (predictable) on the basis of cardiologist Y's ratings (ordinal attribute): The null hypothesis is that this is not true. Because of class sample size imbalance, prior odds weighting is used. Note that, because the class variable and attribute have the same number of categories, and because the hypothesis is directional, LOO analysis is superfluous. Following is the ODA script required to perform this analysis.

```
OPEN ex71.dat;
OUTPUT ex71.OUT;
VARS ratingx ratingy;
CLASS ratingx;
ATTRIBUTE ratingy;
DIRECTIONAL < 1 2 3;
MCARLO ITER 10000;
GO;
```

Compared to conventional approaches, a great advantage of the ODA paradigm is the use of overall PAC as a face valid measure of raw effect strength. Nevertheless, a strong tradition summarizes the reliability of a measure in terms of a theoretical dimension that is bounded between 0 (chance) and 1 (perfect reliability). Fortunately, ODA measures of effect strength—that is, for sensitivity, predictive validity, and total—are bounded between 0 (chance) and 100 (perfect reliability), and therefore are directly analogous to the traditional reliability metric. As in conventional methods, with ODA we obtain an omnibus reliability index: total effect strength. However, in contrast to conventional methods, with ODA we additionally obtain estimates of the reliability of the measure when it is used in a descriptive capacity versus in a prognostic capacity (effect strength for sensitivity versus predictive value, respectively). As seen, the classification performance of the ODA model corresponds to a moderate omnibus inter-rater reliability of .3708, and to moderate values of both descriptive (.375) and prognostic (.3657) reliability: This result was statistically significant (estimated $p < 0.0001$).

Classification Performance Index	Value (%)
Overall classification accuracy	65.00
Sensitivity (normal)	75.00
Sensitivity (possibly abnormal)	50.00
Sensitivity (definitely abnormal)	50.00
Effect strength, sensitivity	37.50
Predictive value (normal)	90.00
Predictive value (possibly abnormal)	33.33
Predictive value (definitely abnormal)	50.00
Effect strength, predictive value	36.57
Total effect strength	37.08

ODA may be used to assess the inter-rater reliability of any binary measure, as well as of any polychotomous or ordinal measure having ten or fewer discrete response categories. Measures having ten or fewer discrete response categories are frequently used in social and medical literatures (and as well in many other fields). For example, common "small-domain" ordinal attributes include Likert-type ratings, and explicitly defined scales such as 1 = "lower class"; 2 = "middle class"; and 3 = "upper class" (see Osgood, Suci, & Tannenbaum, 1957, p. 85; Russell & Bobko, 1992; Schiffman, Reynolds, & Young, 1981, pp. 22–24).

Example 7.2

As another example of inter-rater reliability analysis involving an ordinal attribute, consider Mueser, Sayers, Schooler, Mance, and Haas's (1993) data from a collaborative treatment study of the efficacy of neuroleptic dosage maintenance and family treatment for schizophrenia. Because this research involved quantified observations of patient behaviors, it was necessary to evaluate the inter-rater reliability of numerous ratings. For example, two psychiatrists (A and B) independently rated, via an explicitly defined five-point scale, the same videotaped psychopathology interviews of ten randomly selected patients on the dimension of "unchanging facial expression." On this scale 1 = "behavior not present," and 5 = "behavior present with extreme severity." The data (tabled are the number of patients) were

Rating by Psychiatrist

		B				
		1	2	3	4	5
	1	3				
	2		2			
A	3			1		
	4				1	1
	5					2

The directional hypothesis is that psychiatrist A's and B's ratings are consistent and fall into the main diagonal of the cross-classification table. Psychiatrist A's ratings (five-category class variable) are therefore directly discriminable (predictable) on the basis of psychiatrist B's ratings (ordinal attribute): The null hypothesis is that this is not true. Following is ODA script used to conduct this priors-weighted analysis (note that LOO analysis is superfluous).

```
OPEN ex72.dat;
OUTPUT ex72.OUT;
VARS ratera raterb;
CLASS ratera;
ATTRIBUTE raterb;
DIRECTIONAL < 1 2 3 4 5;
MCARLO ITER 10000;
GO;
```

As seen, the classification performance of the ODA model corresponds to a statistically significant (estimated $p < 0.0001$), very strong omnibus inter-rater reliability (.8958). The model is marginally more reliable in a prognostic (.9167) versus a descriptive (.875) context.

Classification Performance Index	Value (%)
Overall classification accuracy	90.00
Sensitivity (rating = 1, 2, 3, 5)	100.00
Sensitivity (rating = 4)	50.00
Effect strength, sensitivity	87.50
Predictive value (rating = 1, 2, 3, 4)	100.00
Predictive value (rating = 5)	66.67
Effect strength, predictive value	91.67
Total effect strength	89.58

Model sensitivity was 100% for all response levels except for moderate to severe ratings: Inter-rater agreement was less than perfect only for patients rated as manifesting severe (rating = 4) or extreme (rating = 5) unchanging facial expression. These findings suggest that ratings at the low prevalence end of this measure are highly reliable, and indicate that additional rater training, if conducted, should emphasize cases in which the symptom is present.

Parallel Forms Reliability

In classical test theory, parallel forms are alternative equivalent forms of an instrument which, although constituted by different sets of items, measure exactly the same attribute of the subject of measurement. A single person completing a pair of parallel forms should receive identical scores on them, and across a sample of people the forms should receive identical means and variances. Any intraindividual differences that occur in responses to parallel forms are attributed to random error. To estimate parallel forms reliability, the two parallel forms are typically administered to a sample of observations within a given brief time interval—such as two weeks or less (Nunnally, 1964), and the correlation between the two sets of scores (the *equivalence coefficient*) is computed. According to Mike Strube (personal communication, 2000), "the best way to assess this form of reliability is to choose the time period to match the way the measures will be used in application. That way, the same sources of error contribute to the estimation of reliability and to the measures as they are actually used to assess, for example, change due to treatment." The equivalence coefficient—typically kappa (Cohen, 1960) or the Pearson product-moment correlation coefficient (Royeen, 1989; Yarnold & Mueser, 1989; Zegers, 1991), is used as the estimated parallel forms reliability. The square root of the equivalence coefficient estimates the upper bound of the correlation between the instrument and any other measure (Carmines & Zeller, 1979). Many consider parallel forms methodology to be the preferred procedure for assessing the precision of an instrument (cf. Magnusson, 1967).

Example 7.3

As an example of a parallel forms reliability study involving an ordered attribute, consider Matthews, Krantz, Dembroski, and MacDougall's (1982) data concerning the correspondence between two different procedures for assessing Type A behavior (TAB). One procedure, the Structured Interview, is a standardized clinical interview considered the gold standard in the assessment of TAB. The other procedure was a popular self-report questionnaire measure of TAB known as the Jenkins Activity Survey (JAS). Whereas TAB was assessed using a binary metric for the Structured Interview (Type A = 1; Type B = 0), assessments made using the

JAS were ordinal: Scores between 0 and 3 inclusive were dummy-coded as 3; scores of 4 and 5 were coded as 5; scores of 6 and 7 were coded as 7; scores of 8 and 9 were coded as 9; scores of 10 and 11 were coded as 11; and scores between 12 and 16 inclusive were coded as 16 (see Matthews et al., 1982, Sample 2, p. 307). The results of TAB assessments made using both procedures were given for 186 undergraduate males. Serving as an estimate of the parallel forms (or, more defensibly, *alternative forms*) reliability of TAB assessments made using these procedures, Matthews et al. computed the Pearson correlation between the JAS and TAB assessments made using the Structured Interview—coded using a more sensitive, four-point ordinal scale that was not published: $r = 0.31$, $p < 0.001$. This clearly represents a lower bound for the parallel forms reliability of TAB assessment in light of the obvious differences between self- (JAS) and other- (Structured Interview) based assessment methodologies. To analyze these data with ODA, the directional hypothesis is that observations classified as being Type A by the Structured Interview will have greater JAS scores than observations classified as being Type B by the Interview. Weighting by prior odds is used because of the different numbers of As and Bs. Following is the ODA script needed to perform this analysis.

```
OPEN ex73.dat;
OUTPUT ex73.out;
VARS intervew jas;
CLASS intervew;
ATTRIBUTE jas;
DIRECTIONAL < 0 1;
MCARLO ITER 10000;
LOO;
GO;
```

The resulting ODA model was, If JAS ≤ 8, then predict that interview type = A; otherwise, if JAS > 8, then predict that interview type = B. As seen, though statistically significant (estimated $p < 0.001$), the training classification performance of this model was weak in practical terms (total effect strength = 23.9%), although stable in LOO validity analysis.

Classification Performance Index	Value (%)
Overall classification accuracy	61.83
Sensitivity (Type B)	65.08
Sensitivity (Type A)	60.16
Effect strength, sensitivity	25.24
Predictive value (Type B)	45.56
Predictive value (Type A)	77.08
Effect strength, predictive value	22.64
Total effect strength	23.94

Thus, although there is statistically significant evidence for the parallel forms reliability of these measures of TAB, the effect strength is weak.

Split-Half Reliability

A limitation of both the parallel forms and test-retest (discussed below) methodologies for estimating reliability is that they are infeasible unless the identical sample of observations is available for two separate testing sessions. An advantage of the split-half method is that only one test administration is necessary. The most commonly used procedure for obtaining split-halves is known as the adjusted split-half method. In this procedure, after obtaining a sample of observations who completed the test, the items constituting the test are sorted in order of descending variance. One split-half is then composed of the even-numbered items in the sorted list, and the other split-half is composed of the odd-numbered items. Items are then interchanged between split-halves until they are as similar as possible (this is done in an attempt to meet the requirements of parallel forms). Once the split-halves are constructed, a Pearson correlation coefficient is computed between the two split-half scores, and then corrected for attenuation using the Spearman-Brown prophecy formula (Brown, 1910; Spearman, 1910). The result is interpreted as the split-half reliability for the total test (e.g., see Guttman, 1945; Lyerly, 1958; Magnusson, 1967; Rulon, 1939; Yarnold, 1984).

Example 7.4

To illustrate the evaluation of split-half reliability for an application involving a polychotomous attribute, consider data from research concerning the assessment of psychological androgyny (P. R. Yarnold, 1990, 1994). A total of 68 male undergraduates completed a self-report survey androgyny measure assessing two dimensions—instrumentality (I) and expressiveness (E), each using 20 items. In order to form split-halves, the 20 "I" items were randomly divided into two split-halves (I_1 and I_2), and so were the 20 "E" items (E_1 and E_2). Then, separately for each pair of corresponding split-halves—that is, for (I_1, E_1) and again for (I_2, E_2), each undergraduate was classified into one of four mutually exclusive and exhaustive polychotomous categories reflecting conceptually distinct sex-role types: androgynous (dummy-coded using 1); instrumentally typed (2); expressively typed (3); and undifferentiated (4). Finally, a 4×4 contingency table was created by crossing type as assigned by the first split-half (rows) versus type as assigned by the second split-half (columns).

Type by Split-Half	#2			
#1	1	2	3	4
1	11	2	2	0
2	3	11	0	4
3	2	0	13	4
4	1	5	1	9

To evaluate the split-half reliability of this four-category assessment procedure via ODA, these data were analyzed under the directional alternative hypothesis that observations classified as being type t (t = 1, 2, 3, or 4) by split-half 1 (four-category class variable) would be similarly classified as type t by split-half 2 (polychotomous attribute). Prior odds weighting is specified because the class category sample sizes are imbalanced. LOO analysis is superfluous because the class variable and attribute have the same number of categories, and a directional hypothesis is used. The ODA script required to conduct this analysis follows.

```
OPEN ex74.dat;
OUTPUT ex74.out;
TABLE 4;
CATEGORICAL ON;
CLASS ROW;
DIRECTIONAL < 1 2 3 4;
MCARLO ITER 10000;
GO;
```

As seen, the classification performance of the ODA model corresponds to a statistically significant (estimated p < 0.0001), relatively strong omnibus inter-rater reliability (.5319). The model is comparably reliable when used in a prognostic (.5334) versus a descriptive (.5304) capacity. Note also that the model is robust, with all sensitivities, predictive values, and PAC and effect strength indices exceeding 50%.

Classification Performance Index	Value (%)
Overall classification accuracy	64.71
Sensitivity (type = 1)	73.33
Sensitivity (type = 2)	61.11
Sensitivity (type = 3)	68.42
Sensitivity (type = 4)	56.25
Effect strength, sensitivity	53.04
Predictive value (type = 1)	64.71
Predictive value (type = 2)	61.11
Predictive value (type = 3)	81.25
Predictive value (type = 4)	52.94
Effect strength, predictive value	53.34
Total effect strength	53.19

Of course, as is true for all the examples we discuss, although the present procedure is illustrated using an application taken from the social sciences, it is also appropriate for evaluating the split-half reliability of any measurement methodology—irrespective of substantive context—so long as the classifications made by that methodology are into mutually exclusive categories, or "types."

Temporal Reliability

The usefulness of a measurement methodology is dependent not only on its validity and reliability, but also on its responsiveness, defined in terms of its ability to detect minimal yet clinically important differences that occur over time or treatments (Guyatt, Walter, & Norman, 1987; Shea, Norcini, Baranowski, Langdon, & Popp, 1992). Thus, it is quite natural to think of an instrument as being reliable even though it may give different results for a single observation measured at two or more points in time (Magnusson, 1967). Low responsivity can be induced by observer biases that serve to reduce the accuracy of ratings, such as halo biases (ratings that are consistent across—and fail to distinguish among—the evaluation dimensions), leniency and severity biases (ratings that are higher or lower than warranted by actual performance, respectively), and central tendency and range restriction biases (failure to use the full range of the rating scale) (e.g., see Kahneman, Slovic, & Tversky, 1982; Saal, Downey, & Lahey, 1980). It is also important to consider the possibility of reactivity bias induced by the measurement methodology; that is, bias whereby the act of measuring a phenomenon induces changes in the state of the objects of measurement (Nunnally, 1978).

These issues concerning the ability of an instrument to measure change over time notwithstanding, it is also reasonable (and a very common practice) to twice administer a single test to a single sample of observations and to treat the correlation between the two sets of scores as the test–retest or temporal reliability of the instrument (e.g., Yarnold, Mueser, Grau, & Grimm, 1986; Yarnold & Mueser, 1989; Yarnold, 1992). It is presumed that responses to the test will remain correlated over time because they measure the same true score (e.g., Carmines & Zeller, 1979). Though very popular in a variety of disciplines, the test–retest procedure has been criticized on both statistical (lack of independent groups) and methodological (the items constituting the test reflect only one sample from a population of items; memory confounds measurement) grounds (Carmines & Zeller, 1979; Royeen, 1989).

Using ODA, it is a straightforward procedure to evaluate the test–retest reliability of an instrument. Subsequent chapters discuss alternative methodologies for assessing temporal and sequential processes, including Markov models, the intraclass reliability coefficient, the autocorrelation function, the turnover table approach, and single-subject analysis.

Example 7.5

To illustrate the evaluation of test–retest reliability for an application involving an ordered attribute having ten or fewer response levels, consider data from Bryant and Yarnold (1990) concerning the temporal stability of affective (emotional) experience. Participants were 160 undergraduates who, with a two-week retest interval separating the testings, twice completed a self-report survey assessing a variety of emotions. Survey items are single-word descriptors of different types of affective experience. People completing the survey indicate, using a 5-point Likert scale, the degree to which the affect described by each item constitutes an accurate description of their current state of mind (0 = *not at all accurate*, 4 = *very accurate*).

For the purposes of this example, four items were selected for analysis: peeved, lonely, cheerful, and friendly. Notice that the first and second items reflect negative affects, and that the third and fourth items reflect positive affects. Also, notice that the first and third items reflect relatively temporary and rapidly changing emotions (emotional states), whereas the second and fourth items reflect relatively stable and slowly changing dispositions (personality traits).

Because undergraduates received two scores on each of these four items, the last character of the ODA script name for each item is an integer—either 1 or 2—which indicates that the score is from the first or second testing, respectively. The data file also includes information on gender (female = 0, male = 1), but this was not used in this example. The DIRECTIONAL command reflects the directional hypothesis that responses to these items should be directly consistent between testings. Observations on an item for the second testing was treated as the five-category class variable, and the corresponding score from the first testing was treated as the polychotomous attribute. The MCARLO command reflects a Sidak criterion for four tests of statistical hypotheses. LOO analysis is superfluous because the class variable and attribute have the same number of categories, and a directional hypothesis is specified. Weighting by prior odds is used because of the class category sample-size imbalance. The ODA script used to accomplish these analyses follows.

```
OPEN ex75.dat;
OUTPUT ex76.out;
VARS gender peeved1 peeved2 lonely1 lonely2 cheer1 cheer2 friend1 friend2;
DIRECTIONAL < 0 1 2 3 4;
MCARLO ITER 10000 TARGET .05 SIDAK 4;
CLASS peeved2;
ATTR peeved1;
TITLE test-retest reliability for peeved;
GO;
CLASS lonely2;
ATTR lonely1;
TITLE test-retest reliability for lonely;
GO;
CLASS cheer2;
ATTR cheer1;
TITLE test-retest reliability for cheer;
CLASS friend2;
ATTR friend1;
TITLE test-retest reliability for friend;
GO;
```

Consider first the findings for the two trait-like enduring dispositions—lonely and friendly. As seen, classification performance was of moderate practical significance and was statistically significant: For both models, estimated $p < 0.0001$, confidence for $p < 0.01275 = 99.99\%$. Note that model sensitivity was highest at the extreme poles (i.e., codes 0 and 4) of the class variable. These results are consistent with the a priori hypothesis concerning the temporal stability of these two trait-like dispositions, although the practical significance of these effects is moderate.

Classification Performance Index	Lonely	Friendly
Overall classification accuracy	59.38	40.00
Sensitivity (rating = 0)	77.11	45.16
Sensitivity (rating = 1)	42.86	35.14
Sensitivity (rating = 2)	35.00	36.84
Sensitivity (rating = 3)	27.27	35.71
Sensitivity (rating = 4)	75.00	66.67
Effect strength, sensitivity	39.31	29.88
Predictive value (rating = 0)	77.11	50.00
Predictive value (rating = 1)	40.91	38.24
Predictive value (rating = 2)	36.84	31.82
Predictive value (rating = 3)	30.00	50.00
Predictive value (rating = 4)	75.00	33.33
Effect strength, predictive value	39.96	25.85
Total effect strength	39.64	27.86

In contrast to the findings for the trait-like dispositions, no ODA model was found for either of the two state-like (transient) affects—peeved and cheerful. This implies that, for these two items, at least one of the class levels is empty (e.g., no undergraduate answered these items using one of the five possible response categories), or that the directional hypothesis is untenable in light of the temporal structure actually underlying the data. Examination of the data reveals that none of the classes are empty. Thus, the directional hypothesis is untenable.

Nonlinear Reliability

Clearly, the linear reliability hypothesis—which stipulates that data will fall in the major diagonal of the reliability cross-classification table—is not always an appropriate representation of one's data. If such a linear directional hypothesis fails to find support, does this mean that no reliable process underlies the data—that the responses reflect only random error? Or is it possible that a reliable nonlinear pattern underlies the data?

In fact, with ODA it is not uncommon to identify reliable nonlinear models that explain data in reliability contingency tables. As an illustration of this, we continue the prior example in which the linear hypothesis did not find support. The peeved and cheerful data were reanalyzed, setting DIRECTIONAL to OFF, and reparameterizing the Sidak target for six tests of statistical hypotheses. These analyses were accomplished using the following ODA script.

```
DIRECTIONAL OFF;
MCARLO ITER 10000 TARGET .05 SIDAK 6;
LOO;
CLASS peeved2;
ATTR peeved1;
TITLE exploratory temporal analysis for peeved;
GO;
CLASS cheer2;
ATTR cheer1;
TITLE exploratory temporal analysis for cheer;
GO;
```

Considering first the results for peeved, the nondirectional ODA model yielded weak training classification performance (total effect strength = 21.83%), which diminished in LOO validity analysis (total effect strength = 8.03%). Furthermore, the training performance of the model failed to achieve either the experimentwise or the generalized criterion for statistical significance: estimated $p < 0.071$. It is therefore concluded that these data are unreliable.

Considering cheerful next, the nondirectional ODA model achieved was practically weak (total effect strength = 23.2%), but statistically significant (estimated $p < 0.0005$, confidence for $p < 0.00851 = 99.99\%$) classification performance, that decreased marginally in LOO analysis (total effect strength = 17.32%). The temporal structure identified by the nondirectional ODA model for cheerful is largely inconsistent with the temporal structure implied by the original directional hypothesis. The original directional model may be summarized symbolically as

$0 \rightarrow 0, 1 \rightarrow 1, 2 \rightarrow 2, 3 \rightarrow 3$, and $4 \rightarrow 4$,

where, for example, "$0 \rightarrow 0$" means that a score of 0 at the first testing is related to a score of 0 at the second testing. In contrast, the nondirectional model is clearly nonlinear for low scores:

$0 \rightarrow 2, 1 \rightarrow 0, 2 \rightarrow 1, 3 \rightarrow 3$, and $4 \rightarrow 4$,

where, for example, "$0 \rightarrow 2$" means that a score of 0 at the first testing is related to a score of 2 at the second testing. Comparison of these two structures reveals that only the behavior of response categories 3 and 4 are consistent.

Although the structure identified by the nondirectional ODA model is inconsistent with the structure originally used to define temporal reliability, it nonetheless represents a *reliable nonlinear temporal pattern* underlying the data. According to this nonlinear structure, when

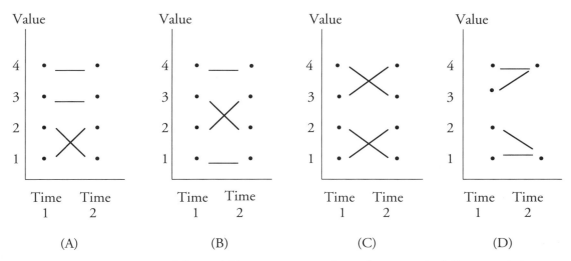

Figure 7.1. Examples of four different types of nonlinear reliability models.

assessed across testings, people who are rarely cheerful (Attribute Category 0) tend to become occasionally cheerful (Class Category 2); people who are infrequently cheerful (Attribute Category 1) tend to become rarely cheerful (Class Category 0); people who are occasionally cheerful (Attribute Category 2) tend to become infrequently cheerful (Class Category 1); and people who are usually cheerful (Attribute and Class Category 3) or almost always cheerful (Attribute and Class Category 4) tend to remain as such. These results suggest that cheerful affect is stable at higher, frequently exhibited levels, but it shows regression at lower, infrequently exhibited levels.

Indeed, several different types of nonlinear reliability models are possible in the ODA paradigm (four types are discussed below). It is possible that any one of these types of nonlinear models underlies the data, regardless of whether temporal, inter-rater, split-half, or parallel forms reliability is being assessed. These models may be specified in an a priori manner (using ODA software, via the DIRECTIONAL command) or they may emerge in exploratory analysis. For example, in the first type of nonlinear model, the attribute may be reliable (stable) at higher values but show regression at lower levels (A in Figure 7.1). The model describing the temporal structure of cheerful affect is this type of nonlinear model. Of course, the symmetric model—in which the attribute is reliable at lower levels but shows regression at higher levels—is also possible.

In the second type of nonlinear model, the attribute may be reliable at extreme values, but show regression at intermediate values (B in Figure 7.1). This type of model is often appropriate if one constructs a class variable or an attribute using a mean- or median-split procedure, in which observations are assigned into different categories on the basis of their location relative to the mean, median, or some other cutpoint(s) on a continuum. Typically, greatest instability occurs near the cutpoint (in 7.1B, the cutpoint might be at 2.5, for example).

In the third type of nonlinear model, the attribute may show local regression throughout its range (C in Figure 7.1). Although scores tend to be positively related over testings, parallel forms, split-halves, or raters—that is, lower scores (1 and 2) at the first testing are associated with lower scores at the second testing, and higher scores (3 and 4) at the first testing are associated with higher scores at the second testing—there is, nevertheless, evidence of local instability in the data.

Finally, the fourth type of nonlinear model is a degenerate nonlinear solution (D in Figure 7.1). Note that no observations were classified into response categories 2 or 3 at the second testing. By definition, model degeneracy exists when there is at least one such missing response category value. This particular example of a degenerate model illustrates a polarization phenomenon, in which extreme scores at the first testing remain stable, and intermediate scores at the first testing become more extreme (polarized) at the second testing (or vice versa).

Intraclass Correlation

The intraclass correlation coefficient of reliability, R, is a widely used statistic that is based on a special-purpose analysis of variance (ANOVA) that derives from the classical test theory notion of reliability, and is the focus of generalizability theory (e.g., Brennan, 1983; Cronbach Rajaratnam, & Gleser, 1963; Fleiss, 1986; Strube, 2000). Briefly, in this approach it is assumed that, for any measurement, there is an underlying error-free true score that cannot be directly observed. Furthermore, it is assumed that an observed score is equal to the true score plus an independent random error score, and the variance of the observed score is equal to the sum of the variances of the true and error scores. R is equal to the ratio of the variance of the true score divided by the variance of the observed score, and is interpreted as reflecting the proportion of the variance of a measure that is attributable to interobservation variation in error-free scores. R approaches its theoretical maximum value (1) as the variance of the error score approaches zero and approaches its theoretical minimum value (0) as the variance of the error score approaches infinity.

Generalizability theory methodology is too complex to review adequately here, but a simple application—known as a fixed effect reliability study (FERS)—provides a sense of what is involved and how ODA may be useful. In the typical data configuration corresponding to a FERS, rows correspond to observations and columns to raters, and the data are raters' evaluations of the observations (each rater evaluates each observation). ANOVA for this FERS produces two F statistics: one for observations (which is of little theoretical value beyond its effect on the error term and therefore is not evaluated), and one for raters (the null hypothesis is that observations' mean ratings are the same across raters). Rejection of the null hypothesis for the main effect of raters implies that raters provided dissimilar ratings for observations—that is, *differential measurement bias* exists (the precise nature of the inter-rater differences requires determination). The extent to which a sample of raters produce consistent ratings is assessed by computing R. When inter-rater differences are small relative to interobservation differences, it is not uncommon to obtain a significant main effect for raters—indicating the presence of significant differential measurement bias, and yet to obtain an R that is near its theoretical upper limit. Ambiguous findings are possible as the magnitude of inter-rater differences increases relative to the magnitude of interobservation differences and the maximum attainable level of R diminishes (cf. Fleiss, 1986; Guyatt, Walter, & Norman, 1987).

ODA may be used to analyze data from a FERS by configuring rater as the class variable: The null hypothesis is that the attribute (raters' evaluations of observations) cannot be used to successfully discriminate raters. Rejection of this null hypothesis implies that raters did not assign observations the same scores and thus that significant differential measurement bias exists (i.e., ratings are not strictly reliable). Failure to reject the null hypothesis implies that raters could not be discriminated on the basis of their ratings of observations and is not inconsistent with the hypothesis that the raters are reliable. Of course, the finding that significant differential measurement bias does not exist does not imply that the data are reliable: Inter-rater reliability must still be assessed. Finally, whereas ANOVA requires that data conform to many often untenable assumptions for p to be meaningful, the p obtained by ODA does not depend on distributional assumptions.

Example 7.6

Fleiss (1986) provides an example of a FERS for which the objective was to evaluate the reliability of four dentists' ratings of the state of ten patients' teeth. The same four dentists independently examined the permanent teeth of each of ten patients and recorded the number of decayed, missing, and filled surfaces observed (DMFS score). ANOVA was used with rater treated as a fixed effect, and revealed a significant rater effect ($p < 0.001$). This finding indicates that the raters differed systematically from each other in terms of the mean DMFS ratings they assigned to patients: evidence of significant differential measurement bias. Inconsistent with this finding, $R = .92$, indicative of very strong reliability.

These data (ex76.dat) were analyzed via ODA: Dentist (DENTIST; dummy-coded as 1 through 4) constituted the class variable, and DMFS score (DMFS) constituted the attribute. Priors-weighting is not needed because each dentist rated the same set (i.e., the same number) of patients. The ODA script for this analysis is

```
OPEN ex76.dat;
OUTPUT ex76.out;
VARS dentist dmfs;
CLASS dentist;
ATTRIBUTE dmfs;
PRIORS OFF;
LOO;
MCARLO ITER 10000;
GO;
```

The classification performance achieved by the ODA model in training analysis was weak (total effect strength = 17.6%) and not statistically significant (estimated $p < 0.99$), and the classification performance collapsed in LOO validity analysis (total effect strength = −27.2%). These results strongly suggest that dentists could not be discriminated from each other on the basis of the DMFS scores they assigned to patients. These results contradict the ANOVA finding of a significant rater effect and are not inconsistent with the hypothesis that ratings are strictly reliable. Because there was no evidence of inter-rater discriminability on the basis of ratings made of observations, raw ratings may be used in an ODA-based analysis of the strict reliability of these data. However, if instead the inter-rater effect had been found to be statistically significant, then data would first be normatively standardized separately by rater and then ODA would be used to test the relative reliability hypothesis for these data.

CHAPTER 8
Validity Analysis

*I*n this manuscript we use two meanings for the term *validity*. The first use refers to the *legitimacy of the estimated classification performance* for an ODA model. That is, to what extent is the percentage accuracy in classification (PAC) achieved by the model—which is developed using a specific training sample—indicative or representative of the PAC that would be achieved by subsequently using that model to make classifications with independent samples? ODA software provides several techniques for assessing the validity of estimated PAC, two of which—hold-out (discussed below) and LOO—are widely used in this context. Subsequent chapters discuss procedures that may be used to evaluate the generalizability of an ODA model when it is simultaneously used with two or more samples; to split large samples and identify an ODA model providing a more realistic estimate of PAC than would be obtained if the sample had instead been analyzed as a single group; and to optimize the cross-validated classification performance achieved by suboptimal multiattribute methods such as logistic regression analysis or Fisher's discriminant analysis.

The second use of the term *validity*—which we address after discussing hold-out validity, is drawn from the area of psychometrics (e.g., Allen & Yen, 1979; Bryant, 2000; Magnusson, 1967; Nunnally, 1978). A classic and commonly used definition of validity refers to the extent to which a measuring device "measures what it is intended to measure." However, according to Dr. Mike Strube (personal communication, 2000), another perhaps more intuitive definition might be "the appropriateness of the label attached to a measure. A good example is head circumference as measured by a tape measure. If I label this as a measure of hat size, no one will dispute the validity of that assertion. But, if I label the same instrument as a measure of intelligence (under the misguided assumption that head size is a good proxy for brain size, which is a good proxy for intelligence), then many will dispute the validity of the assertion."

Hold-Out (Cross-Generalizability) Validity

Discussed earlier, the most straightforward method of assessing the classification error rate of an ODA model is known as the hold-out, cross-generalizability, or cross-validation procedure. The hold-out validity procedure essentially involves attempting to replicate one's original findings using an independent random sample. In order to estimate the hold-out validity of an ODA model (a) first develop the model using a training sample and then (b) use the model to classify one or more independent hold-out samples. The classification error rate observed for the hold-out sample(s) is then used as the estimated hold-out classification error rate for the model (e.g., Geisser, 1975; Stone, 1974).

Example 8.1

As an example of a hold-out validity analysis for an application involving a multicategory class variable and a polychotomous attribute, consider Mosteller's (1968) data on fathers' and sons' occupations in England and in Denmark. Occupations were classified into the same five categories—dummy-coded using integers between 1 and 5—for fathers and sons from both countries. Separately for both countries, a cross-classification table was created with fathers' occupation constituting the rows, and sons' occupation constituting the columns (the British table is ex81a.dat, and the Danish table data is ex81b.dat).

British Father	Son					Danish Father	Son				
	1	2	3	4	5		1	2	3	4	5
1	50	45	8	18	8	1	18	17	16	4	2
2	28	174	84	154	55	2	24	105	109	59	21
3	11	78	110	223	96	3	23	84	289	217	95
4	14	150	185	714	447	4	8	49	175	348	198
5	3	42	72	320	411	5	6	8	69	201	246

The directional alternative hypothesis is that fathers' and sons' occupational classes tend to be the same, such that son's occupational code (five-category class variable) should be directly discriminable on the basis of father's occupational code (polychotomous attribute): The null hypothesis is that this is not true. British data are used for training analysis, and Danish data are used for hold-out analysis. Note that the decision concerning which sample to use for training purposes and which to use for hold-out purposes is arbitrary in this example, because there is a directional hypothesis and the number of class variable and attribute categories is identical. The Sidak criterion in the MCARLO command is appropriate for experimentwise

$p < 0.05$ and two tests of statistical hypotheses—one for training results, and one for hold-out results. Finally, weighting by prior odds is used because of class category sample size imbalance in both data sets. Following is the ODA script for this analysis.

```
OPEN ex81a.dat;
OUTPUT ex81a.out;
CATEGORICAL ON;
TABLE 5;
CLASS COL;
DIRECTIONAL < 1 2 3 4 5;
MCARLO ITER 10000 TARGET .05 SIDAK 2;
HOLDOUT ex81b.dat;
TITLE Danish hold-out sample;
GO;
OPEN ex81b.dat;
HOLDOUT ex81b.dat;
TITLE British hold-out sample;
GO;
```

The training classification performance achieved using the directional ODA model with British data was weak (total effect strength = 23.5%), yet it was statistically significant because of the relatively large sample (estimated $p < 0.0001$; confidence > 99.99% for $p < 0.02533$).

Classification Performance Index	British	Danish
Overall classification accuracy	41.69	42.07
Sensitivity (code = 1)	47.17	22.78
Sensitivity (code = 2)	35.58	39.92
Sensitivity (code = 3)	23.97	43.92
Sensitivity (code = 4)	49.97	41.98
Sensitivity (code = 5)	40.41	43.77
Effect strength, sensitivity	24.27	23.10
Predictive value (code = 1)	38.76	31.58
Predictive value (code = 2)	35.15	33.02
Predictive value (code = 3)	21.24	40.82
Predictive value (code = 4)	47.28	44.73
Predictive value (code = 5)	48.47	46.42
Effect strength, predictive value	22.72	24.14
Total effect strength	23.50	23.62

Classification performance achieved by the directional ODA model used with Danish (hold-out) data was comparably weak (total effect strength = 23.62). How does one determine p for the hold-out results? Simply use ODA to determine p for the priors-weighted directional alternative hypothesis that data will fall in the major diagonal of the hold-out table. This is accomplished using the following ODA script (entered directly in free format for expository purposes, data were taken directly from the output for hold-out results from the prior analysis).

```
OPEN DATA;
OUTPUT ex81.out;
CATEGORICAL ON;
TABLE 5;
CLASS ROW;
DIRECTIONAL < 1 2 3 4 5;
MCARLO ITER 10000 TARGET .05 SIDAK 2;
DATA;
18 24 23 8 6
17 105 84 49 8
16 109 289 175 69
4 59 217 348 201
2 21 95 198 246
END DATA;
GO;
```

Thus, hold-out classification performance was statistically significant with experimentwise $p < 0.05$ (estimated $p < 0.0001$; confidence > 99.99% for $p < 0.02533$).

As is true for reliability, tradition involves the use of a *validity coefficient*, also known as a *replicability coefficient*—that is bounded between 0 (chance) and 1 (perfect replication)—as a summary estimate of the hold-out validity of a measure. Accordingly, as is true for reliability, for ODA analyses there are three such standardized validity coefficients: the validity coefficients for the model when used in a descriptive versus prognostic capacity (indicated by the effect strength for sensitivity and predictive value, respectively), and the omnibus validity coefficient for the model considered from both perspectives (indicated by the total effect strength). To obtain these coefficients, simply move the decimal point in the effect strength index two spaces to the left (i.e., divide the effect strength index by 100).

For the present example, for the Danish data, the three hold-out validity coefficients are: descriptive validity coefficient = .231; prognostic validity coefficient = .241; and omnibus validity coefficient = .236. Had the British data been treated as the hold-out sample rather than the Danish data, then the three hold-out validity coefficients would have been descriptive validity coefficient = .243; prognostic validity coefficient = .227; and omnibus validity coefficient = .235.

Example 8.2

As an example of a hold-out validity analysis for an application involving a multicategory class variable and an interval attribute, consider Yarnold's (1987) data on Type A behavior (TAB) and psychological instrumentality. Five independent random samples of college undergraduates completed two surveys: The first assessed TAB using a 22-point interval scale. A popular procedure for defining behavioral medicine typologies—the class variable—with this measure (as is true for many such measures) involves trichotomizing the 22-point scale: here, score ≤ 5 = Type B (dummy-coded as 1); $5 <$ score ≤ 10 = Type X (coded as 2); and score > 10 = Type A (coded as 3). The second survey assessed instrumentality with 20 items, each answered using a 7-point scale: This interval attribute has 140 possible response levels. According to theory, Type As should score the highest scores on this attribute; Type Bs should score lowest; and the scores of Type Xs should be intermediate. To begin the analysis, the first data set (Example 82a) was randomly selected as the training sample: The remaining four data sets thus constituted hold-out validity samples. Because of the different numbers of As, Bs, and Xs, prior odds weighting was used. The Sidak criterion is appropriate for five tests of statistical hypothesis involving the evaluation of classification performance: one test for the training analysis, and one test each for hold-out results. Note that Monte Carlo simulation and LOO analysis were turned OFF after the first hold-out run: Because these analyses are conducted for the training sample, conducting them more than once would be superfluous. Following is the ODA script used to perform these analyses.

```
OPEN ex82a.dat;
OUTPUT ex82.out;
VARS tabtype tabscore instru;
CLASS tabtype;
ATTRIBUTE instru;
DIRECTIONAL < 1 2 3;
MCARLO ITER 10000 TARGET .05 SIDAK 5;
LOO;
HOLDOUT ex82b.dat;
TITLE hold-out for sample b;
GO;
MCARLO OFF;
LOO OFF;
```

```
HOLDOUT ex82c.dat;
TITLE hold-out for sample c;
GO;
HOLDOUT ex82d.dat;
TITLE hold-out for sample d;
GO;
HOLDOUT ex82e.dat;
TITLE hold-out for sample e;
GO;
```

The ODA model identified in the training analysis was: If instru ≤ 93.5, then predict Type B; otherwise, if 93.5 < instru ≤ 107.5, then predict Type X; otherwise, if instru > 107.5, then predict Type A. Training classification performance was statistically significant (estimated $p < 0.0001$) but moderate in practical terms (total effect strength = 26.88%), and performance fell in LOO validity analysis (total effect strength = 20.89%). For this ODA model, the classification performance—both in LOO validity analysis for the training sample, and when applied to the four hold-out samples—is summarized below (truncated at the first decimal).

Classification Performance Index	Training (LOO)	Hold-Out Sample			
		B	C	D	E
Overall classification accuracy	42.8	45.1	46.9	50.9	51.1
Sensitivity (Type A)	64.5	70.2	60.3	63.2	64.9
Sensitivity (Type X)	28.7	39.0	38.7	44.8	41.8
Sensitivity (Type B)	55.3	41.4	53.7	59.2	60.3
Effect strength, sensitivity	24.3	25.3	26.4	33.7	33.5
Predictive Value (Type A)	45.4	37.5	38.0	35.2	42.0
Predictive Value (Type X)	58.6	59.6	60.1	69.9	62.6
Predictive Value (Type B)	30.7	35.8	41.3	41.0	46.9
Effect strength, predictive value	17.4	16.4	19.7	23.0	25.7
Effect strength, total	20.9	20.9	23.0	28.3	29.6

Eyeball inspection suggests that classification performance was relatively consistent across samples. As in the preceding example, supplemental analyses were undertaken to determine p for each of the hold-out samples: This was accomplished using the following ODA script.

```
OPEN DATA;
OUTPUT ex82new.out;
CATEGORICAL ON;
```

```
TABLE 3;
CLASS ROW;
DIRECTIONAL < 1 2 3;
MCARLO ITER 10000 TARGET .05 SIDAK 5;
TITLE p for hold-out sample b;
DATA;
29 31 10
47 59 45
5 9 33
END DATA;
GO;
TITLE p for hold-out sample c;
DATA;
36 30 11
47 62 51
4 21 38
END DATA;
GO;
TITLE p for hold-out sample d;
DATA;
32 19 3
43 79 54
3 15 31
END DATA;
GO;
TITLE p for hold-out sample e;
DATA;
38 20 5
40 62 46
3 17 37
END DATA;
GO;
```

The results of these analyses revealed that results for all four hold-out validity analyses were statistically significant at experimentwise $p < 0.05$. Considered as a whole, these findings support the a priori hypothesis concerning the relationship between TAB and instrumentality and indicate that the weak relationship is cross-generalizable.

Construct Validity

Considered by many to be the most important and useful form of the various types of psychometric validity, construct validity is concerned with the consistency and strength with which a given measure relates to other measures to which, according to theory, it should relate (e.g., see Carmines & Zeller, 1979; Cook & Campbell, 1978; Cronbach & Meehl, 1955; Lord & Novick, 1968; Yarnold & Bryant, 1988). As described by Magnusson (1967),

> We begin from a logically defined variable. . . . This variable is included as a logical construct in a system of constructs, in which all of the concepts logically belong, and where all the relationships are explained by a theory. From this theory certain practical consequences can be derived about the outcome of the test under certain conditions. These consequences can be tested. If the result is what was expected in a series of such tests, the test is said to have construct validity for the variable tested. (p. 130)

Example 8.3

As an example of a construct validity analysis for an application involving a binary class variable and attribute, consider Jenkins, Zyzanski, Ryan, Flessas, and Tannenbaum's (1977) data on Type A behavior (TAB) and coronary artery disease (CAD). According to theory, because TAB is predictive of cardiovascular disease and because CAD is a form of cardiovascular disease, TAB should thus be predictive of CAD. To test the validity of the TAB construct, Jenkins et al. surveyed male patients undergoing coronary angiography. On the basis of their scores on a survey, patients were classified as being either Type A or Type B. In addition, patients were independently classified as having severe (two or more coronary arteries obstructed by at least 50%) or mild (one or no coronary arteries obstructed by at least 50%) CAD. For ODA, the directional alternative hypothesis is that, compared to Type Bs, Type As (binary class variable) should have more severe CAD (binary attribute): The null hypothesis is that A/B Type cannot be discriminated on the basis of CAD severity.

A/B Type	Mild CAD	Severe CAD
Type A	12	33
Type B	24	19

Data were entered in tabular format: TAB was the rows (row 1 = A, row 2 = B), and CAD was the columns (column 1 = mild, column 2 = severe). It is hypothesized that As (code = 1) should have more severe CAD (code = 2). Thus, lower scores on the class variable should be associated with higher scores on the attribute. Because of class category sample imbalance (although minor), prior odds weighting is used. LOO validity analysis would be superfluous. Following are the ODA script used to conduct this analysis and the classification performance summary.

```
OPEN ex83.dat;
OUTPUT ex83.out;
TABLE 2;
CATEGORICAL ON;
CLASS ROW;
DIRECTIONAL > 1 2;
MCARLO ITER 10000;
GO;
```

Classification Performance Index	Value (%)
Overall classification accuracy	64.77
Sensitivity (Type A)	73.33
Sensitivity (Type B)	55.81
Effect strength, sensitivity	29.15
Predictive value (Type A)	63.46
Predictive value (Type B)	66.67
Effect strength, predictive value	30.13
Total effect strength	29.64

As seen, the classification performance of this directional ODA model was moderate in practical terms (omnibus construct validity coefficient = .2964), and was statistically significant (exact $p < 0.005$). These findings are consistent with the a priori hypothesis and offer moderate support for the construct validity of TAB as a precursor of cardiovascular disease endpoints.

Convergent and Discriminant Validity

Campbell and Fiske (1959) describe an experimental methodology—involving the multitrait–multimethod matrix—that represents a comprehensive nomological framework for assessing construct (and other types of) validity. They argue that, in the process of assessing the validity of a construct, although it is important to demonstrate that a measure does relate to other measures to which it theoretically should relate, this is not a sufficient test of validity. Rather, it is also important to demonstrate that a measure does not relate to other measures to which it theoretically should *not* relate—for example, because this latter category of measures falls outside of the system within which validity is being assessed. Investigations involving measures that theoretically should be related are concerned with the assessment of *convergent validity*, whereas investigations involving measures that theoretically should be unrelated are concerned with the assessment of *discriminant validity* (cf. Yarnold & Bryant, 1988).

Example 8.4

As an example of a discriminant validity analysis for an application involving a binary class variable and a categorical ordinal attribute, consider Rappaport, McAnulty, and Brantley's (1988) data on Type A behavior (TAB) and both migraine and tension headache. People who experience migraine headaches are characterized by many of the same behaviors as Type As. Also, migraine headaches constitute a vascular disorder, and TAB is believed to influence vascular processes. In contrast, there is no a priori reason to believe that Type As will experience a disproportionately high rate of tension headaches relative to Type Bs.

For this study of the discriminant validity of the TAB construct, Rappaport et al. recruited 30 tension headache sufferers and 30 migraine headache sufferers (binary class variable). On the basis of their survey scores, participants were classified as Type A (score > 75th percentile; dummy-coded as 1), Type X (score between the 75th and 25th percentile; coded as 2), or Type B (score < 25th percentile; coded as 3): TAB is thus a categorical ordinal attribute, with increasing codes reflecting decreasing levels of TAB. The directional hypothesis is that a larger proportion of migraine headache sufferers compared with tension headache sufferers would be Type A: The null hypothesis is that this is not true.

Headache Type	Type A	Type X	Type B
Migraine	16	9	5
Tension	7	13	10

Data were entered using free format: first the code for TAB (A = 1, X = 2, B = 3), and then the code for type of headache (migraine = 1, tension = 2). Because there were 30 observations in both categories of the class variable, weighting by prior odds was turned OFF. LOO validity analysis was conducted. Following is the ODA script used to perform this analysis.

```
OPEN ex84.dat;
OUTPUT ex84.out;
VARS headache tab;
CLASS headache;
ATTRIBUTE tab;
```

```
PRIORS OFF;
DIRECTIONAL < 1 2;
LOO;
MCARLO ITER 10000;
GO;
```

The ODA model was, If TAB ≤ 1.5 (Type A), then predict that headache = 1 (migraine); otherwise, if 1.5 < TAB (Type B, Type X), then predict that headache = 2 (tension).

Classification Performance Index	Value (%)
Overall classification accuracy	65.00
Sensitivity (migraine)	53.33
Sensitivity (tension)	76.67
Effect strength, sensitivity	30.00
Predictive value (migraine)	69.57
Predictive value (tension)	62.16
Effect strength, predictive value	31.73
Total effect strength	30.86

Classification performance, stable in LOO validity analysis, was moderate in practical terms (overall effect strength = 30.86%) and statistically significant (estimated $p < .025$). These results are consistent with the a priori hypothesis and offer moderate support (omnibus discriminant validity coefficient = .3086) for the discriminant validity of the TAB construct.

Example 8.5

As an example of a convergent validity analysis for an application involving a multicategory class variable and a polychotomous attribute, consider Nishikawa, Kubota, and Ooi's (1983) data concerning two different theoretical approaches for the classification of proteins into types. A total of 325 proteins were independently classified into one of four mutually exclusive and exhaustive types (dummy-coded using 1 – 4) twice—once each on the basis of two different theoretical approaches. One classification approach was based on the biological characteristics of the proteins, and the other was based on their amino acid compositions. Because the four types of proteins identified by these two methods theoretically should be consistent, the directional alternative hypothesis is that the codes indicating the type of protein are consistent between methods across proteins: That is, the two methods demonstrate convergent validity. The null hypothesis is that the codes assigned to proteins by the two methods do not correspond: That is, the two methods fail to demonstrate convergent validity.

| Biological | Amino Acid Approach | | | |
Approach	1	2	3	4
1	98	16	5	3
2	13	50	2	8
3	6	4	23	12
4	7	19	14	45

Data were entered in tabular format (rows were codes for biologically defined types of proteins; columns were codes for amino-acid-composition-defined types of proteins) and analyzed using the following ODA script.

```
OPEN ex85.dat;
OUTPUT ex85.out;
CATEGORICAL ON;
TABLE 4;
CLASS ROW;
DIR < 1 2 3 4;
MC ITER 10000;
GO;
```

As seen, the classification performance achieved by this directional ODA model was relatively strong in practical terms (total effect strength = 51.09%), statistically significant (estimated $p < .0001$), and robust, with all classification performance indices exceeding 50%.

Classification Performance Index	Value (%)
Overall classification accuracy	66.46
Sensitivity (protein type 1)	80.33
Sensitivity (protein type 2)	68.49
Sensitivity (protein type 3)	51.11
Sensitivity (protein type 4)	52.94
Effect strength, sensitivity	50.96
Predictive value (protein type 1)	79.03
Predictive value (protein type 1)	56.18
Predictive value (protein type 1)	52.27
Predictive value (protein type 1)	66.18
Effect strength, predictive value	51.22
Total effect strength	51.09

These results are consistent with the a priori hypothesis, and offer relatively strong support (omnibus convergent validity coefficient = .5109) for the convergent validity of the two different protein-typing methods. Of course, the effect is not perfect, and one may wonder whether the incorrectly classified observations reflect random error or if there may be some underlying organization amongst them that is unrevealed to the primary model. Techniques for answering this question appear in chapter 12. Several additional aspects of ODA, addressed in the following chapters, must be introduced first, however.

CHAPTER 9

Optimizing Suboptimal Multivariable Models

Multiattribute applications involve two or more attributes. By using techniques we have already discussed, ODA can assess the utility of each attribute—considered separately—to discriminate between the different class categories and to predict observations' class category membership status. Using ODA in this manner ensures that we will extract the maximum amount of information (relevant to the task of accurate classification) that it is theoretically possible to extract from any single attribute considered alone. Analysis using ODA to evaluate a single attribute is called univariable ODA: *UniODA* ("you-knee-oh-dah") for short.

To the extent that it is possible to accomplish, decision-makers seek perfect classification accuracy. In this context, it is typically the case that classification performance achieved using a model based on a single attribute is worse than would instead be achieved using a model based on a function of two or more attributes: that is, a *multiattribute classification model*. For a multiattribute problem, the classification model that explicitly yields maximum (non-)weighted training classification accuracy is obtained via multivariable ODA: *MultiODA* ("mull-tee-oh-dah") for short (e.g., see Erenguc & Koehler, 1990; Ibaraki & Muroga, 1970; Joachimsthaler & Stam, 1990; Shautsukova, 1975; Soltysik & Yarnold, 1991, 1993, 1994b; Stam & Jones, 1990; Yarnold, Soltysik, & Martin, 1994; Yarnold et al., 1995). Nonlinear MultiODA models have been developed (Soltysik & Yarnold, 1992a), but most research focuses on simpler linear models. Whereas linear MultiODA problems involving a binary class variable and six or fewer binary attributes may be quickly solved for gargantuan samples (involving billions of observations), problems involving ordered or polychotomous attributes may be computationally intractable for samples consisting of as few as 50 observations.

Accordingly, numerous suboptimal heuristic and linear programming approaches to the statistical classification problem have been developed, that obtain models that tend to maximize (non)weighted percentage accuracy in classification (PAC; e.g., Freed & Glover, 1986; Koehler & Erenguc, 1990; Kolesar & Showers, 1985; Soltysik & Yarnold, 1992b). Even though these procedures can classify observations relatively well, they are suboptimal, and thus theoretically may be improved whenever suboptimal models are identified—that is, for most applications. If the sample is too large to conduct MultiODA, which suboptimal classification methodology should one select to tend to maximize (non)weighted PAC? Unfortunately, we do not yet know the answer to this question. Fortunately, UniODA may be used to *optimize the classification accuracy of any suboptimal multivariable model* having ten or fewer class categories. As is generally true, LOO and hold-out validity analysis should be used to assess model generalizability.

Fisher's linear discriminant analysis (FLDA) and logistic regression analysis (LRA) may be the most popular and widely used suboptimal multivariable classification methods (e.g., Efron, 1975; Grimm & Yarnold, 1995; Halperin, Blackwelder, & Verter, 1972; Kleinbaum, Kupper, & Muller, 1988; Stevens, 1992). When such suboptimal methods are used for two-category classification problems, most provide an equation (*response function*) whereby values on the attributes are combined into a single score (*Y-hat*) for each observation. To make classifications, the Y-hats are compared against a given theoretical cutpoint and direction. For example, for FLDA the model is typically, If Y-hat $\leq$ 0, then predict that class = 0; otherwise, predict that class = 1. For LRA the model is often, If Y-hat $\leq$ 0.5, then predict that class = 0; otherwise, predict that class = 1.

Using a procedure known as *ODA-based optimization*, ODA is applied to the Y-hats obtained via a suboptimal multivariable model to determine a new combination of cutpoint and direction that explicitly maximizes (weighted or nonweighted) training PAC (Yarnold & Soltysik, 1991b). To conduct ODA-based optimization of FLDA, LRA, or *any suboptimal two-category classification model* for training data

1. obtain the suboptimal model;
2. using the suboptimal model, obtain Y-hat for each observation;
3. perform priors-weighted ODA on the Y-hats;
4. use the resulting ODA model (not the suboptimal model) to make classifications and compute training PAC; and
5. conduct LOO validity analysis to evaluate the stability of the ODA (i.e., ODA-optimized suboptimal) model.

If, in addition to the training sample, one or more hold-out validity samples is also available, then

6. use the training suboptimal model to obtain Y-hat for observations in the hold-out sample(s); and
7. use the training ODA model to classify observations in the hold-out sample(s) and compute hold-out validity PAC.

Initial research investigating the utility of ODA-based optimization in improving both training and hold-out classification performance of FLDA and LRA models has been encouraging. For example, Yarnold, Hart, and Soltysik (1994) meta-analyzed the results of ODA-based optimization of FLDA and LRA models for 15 data sets representing a variety of substantive areas and problem configurations (i.e., combinations of sample size, class category sample size balance, number and metric of attributes). ODA-based optimization yielded a mean increase of 5.8% in overall PAC in training analysis, and of 3.7% in overall PAC in hold-out validity analysis. When classification performance was weighted by sample size, these mean improvements were 5.5% and 4.8%, respectively. In the time since the meta-analysis was published, sufficient additional studies have reported the use of ODA-based optimization to warrant an update (e.g., Finn & Stalans, 1996; Stalans & Finn, 1995; Weinfurt, Bryant, & Yarnold, 1994; Yarnold & Bryant, 1994; Yarnold, Martin, Soltysik, & Nightingale, 1993). However, because the more recent studies replicate the earlier-noted improvement in the training and hold-out classification performance achieved by the optimized suboptimal models, instead we turn our attention to illustrating the technique of ODA-based optimization.

Optimizing Fisher's Linear Discriminant Analysis

Fisher's linear discriminant analysis (FLDA) is often used in conjunction with multivariate analysis of variance. FLDA is a very popular procedure, and examples of its use may be found quite easily in a broad range of scientific disciplines. Appropriate for applications involving ordered attributes, FLDA is inappropriate for use with categorical attributes.

Example 9.1

As an example of ODA-based optimization of an FLDA model, consider Yarnold, Bryant, Nightingale, and Martin's (1996) data on sympathy and empathy for physicians versus undergraduates (the binary class variable, called *person* in the ODA script). Physicians (coded as 1) will be discriminated from undergraduates (coded as 0) on the basis of four interval attributes: empathic concern (*emp*), perspective-taking (*per*), fantasy (*fan*), and distress (*dis*).

We used conventional statistics software—any system that provides FLDA will suffice—to perform FLDA and obtain a model for discriminating undergraduates versus physicians on the basis of the four attributes. As seen, this FLDA model achieved training classification performance that was moderate in practical terms.

Classification Performance Index	Value (%)
Overall classification accuracy	68.9
Sensitivity (physician)	67.5
Sensitivity (undergraduate)	70.5
Effect strength, sensitivity	38.1
Predictive value (physician)	73.3
Predictive value (undergraduate)	64.4
Effect strength, predictive value	37.8
Total effect strength	38.0

Most statistical software systems do not yet provide LOO validity analysis for FLDA (or other multivariable) models. If desired, a manual LOO validity analysis could be conducted (this was not done here). This example would require 213 separate FLDA analyses: the data (person, emp, per, fan, and dis) are entered in free format in EX91A.dat.

For the FLDA model, Y-hat (i.e., the discriminant function score) was obtained for each observation. The following ODA script was then used to optimize this FLDA model.

```
OPEN ex91b.dat;
OUTPUT ex91b.out;
VARS person yhat;
CLASS person;
ATTRIBUTE yhat;
LOO;
GO;
```

The optimized FLDA model is, If yhat ≤ .148, then predict that person = undergraduate; otherwise, predict that person = physician (recall that the cutpoint is zero for any non-optimized FLDA model). Following is the classification performance summary for the optimized FLDA model in training and LOO analyses.

Classification Performance Index	Training	LOO
Overall classification accuracy	71.77	70.81
Sensitivity (physician)	78.07	77.19
Sensitivity (undergraduate)	64.21	63.16
Effect strength, sensitivity	42.28	40.35
Predictive value (physician)	72.36	71.54
Predictive value (undergraduate)	70.93	69.77
Effect strength, predictive value	43.29	41.31
Total effect strength	42.78	40.83

ODA-based optimization improved the classification performance of FLDA in training analysis: Total effect strength of the optimized FLDA model in LOO validity analysis (40.83%) reflected a 7.4% improvement versus the total effect strength obtained by the non-optimized FLDA model in training analysis (38.0%). Comparing only training analysis results, the optimized model had 12.6% greater total effect strength (42.78%) than the non-optimized model.

Example 9.2

As another example of ODA-based optimization of an FLDA model, consider Yarnold, Stille, and Martin's (1996) data on aging and functional status. Ambulatory medical patients were scored on five continuous attributes assessing functional status: basic activities of daily life, intermediate activities of daily life, psychological well-being, social activity, and quality of social interactions. These attributes are all interactive, with 0 representing the lowest possible score and 100 representing the highest possible score. FLDA was used to obtain a model for discriminating between geriatric (at least 65 years of age; coded as 1) versus non-geriatric (coded as 0) patients (the class variable, age). Because of imbalanced numbers of (non)geriatric patients, prior odds weighting was used. As seen, this FLDA model achieved training classification performance that was moderate in practical terms.

Classification Performance Index	Value (%)
Overall classification accuracy	73.7
Sensitivity (geriatric)	33.3
Sensitivity (non-geriatric)	93.9
Effect strength, sensitivity	27.3
Predictive value (geriatric)	73.3
Predictive value (non-geriatric)	73.8
Effect strength, predictive value	47.1
Total effect strength	37.2

For the FLDA model, Y-hat (i.e., the discriminant function score) was obtained for each observation. The following ODA script was then used to optimize this FLDA model.

```
OPEN ex92.dat;
OUTPUT ex92.out;
VARS age yhat;
CLASS age;
ATTRIBUTE yhat;
LOO;
GO;
```

The optimized FLDA model is, If yhat ≤ −.495, then predict that age = geriatric; otherwise, predict that age = non-geriatric (the cutpoint is zero for any non-optimized FLDA model). Following is the classification performance summary for the optimized FLDA model in training and LOO analyses.

Classification Performance Index	Training	LOO
Overall classification accuracy	77.78	76.77
Sensitivity (geriatric)	51.52	48.48
Sensitivity (non-geriatric)	90.91	90.91
Effect strength, sensitivity	42.42	39.39
Predictive value (geriatric)	73.91	72.73
Predictive value (non-geriatric)	78.95	77.92
Effect strength, predictive value	52.86	50.65
Total effect strength	47.64	45.02

ODA-based optimization improved the classification performance of FLDA in training analysis: The total effect strength of the optimized FLDA model in LOO validity analysis (45.02%) reflected a 21.0% improvement versus the total effect strength obtained by the non-optimized FLDA model in training analysis (37.2%). Comparing training analysis results, the optimized model had 28.0% greater total effect strength (47.64%) than the non-optimized model.

Optimizing Logistic Regression Analysis

Use of logistic regression analysis (LRA) to solve the statistical classification problem has been increasing steadily over the past two decades, and examples of its use may easily be found in many substantive areas. Reasons frequently cited underlying the decision to use LRA rather than FLDA include the ability of LRA to use ordered and categorical attributes and the greater robustness of LRA (compared to FLDA) for violations of its assumptions (cf. Efron, 1975; Halperin et al., 1972; Titterington et al., 1981).

Example 9.3

As an example of ODA-based optimization of an LRA model, reconsider Yarnold, Bryant, et al.'s (1996) data on sympathy and empathy for physicians versus undergraduates, presented in Example 9.1. As seen below, this LRA model achieved training classification performance that was moderate in practical terms. The ODA script in Example 9.1 was used to optimize this LRA model: The only changes to the script included that "ex93.dat" was substituted in the OPEN command, and "ex93.out" in the OUTPUT command.

Classification Performance Index	Non-Optimized Training	Optimized LOO
Overall classification accuracy	71.77	71.29
Sensitivity (physician)	78.07	78.07
Sensitivity (undergraduate)	64.21	63.16
Effect strength, sensitivity	42.28	41.23
Predictive value (physician)	72.36	71.77
Predictive value (undergraduate)	70.93	70.59
Effect strength, predictive value	43.29	42.36
Total effect strength	42.78	41.80

Note that the LRA model resulted in somewhat greater classification performance than the FLDA model (total effect strength = 38.8% versus 37.2%, respectively). The optimized LRA model is, If yhat ≤ .524, then predict that person = undergraduate; otherwise, predict that person = physician (recall that the cutpoint is .5 for any non-optimized LRA model). ODA-based optimization improved the classification performance of LRA in training analysis: The total effect strength of the optimized LRA model in LOO validity analysis (41.8%) represented a 7.7% improvement versus the total effect strength obtained by the non-optimized LRA model in training analysis (38.8%).

Example 9.4

As an example of ODA-based optimization of an LRA model involving categorical attributes, consider Mausner, Mausner, and Rial's (1967) data concerning physician behavior, personal decision making, and smoking behavior. A total of 157 cigarette smokers were interviewed at their doctor's office, and randomly assigned to either an experimental group (in which the physician advised the person to quit smoking and offered a cigarette substitute; dummy-coded as 1) or a control group (the physician said nothing about smoking; coded as 0). Participants all responded to a survey question concerning whether (coded as 1) or not (coded as 0) they were willing to make a commitment to terminate their smoking. These two variables—group assignment and willingness to commit—are the binary attributes for the LRA. Finally, the class variable (behavior) was a binary indicator of whether (coded as 1, N = 121) or not (coded as 0, N = 36) the person had terminated their smoking when interviewed at a one-week follow-up. LRA was conducted to obtain a model for discriminating between persons who terminated versus continued smoking (behavior) on the basis of their group assignment and willingness to commit. As seen, the non-optimized LRA model achieved training classification performance that was relatively strong in practical terms.

Classification Performance Index	Non-Optimized Training	Optimized LOO
Overall classification accuracy	79.6	91.1
Sensitivity (quitter)	76.2	100.0
Sensitivity (smoker)	81.9	61.1
Effect strength, sensitivity	58.1	61.1
Predictive value (quitter)	73.8	89.6
Predictive value (smoker)	83.7	100.0
Effect strength, predictive value	57.6	89.6
Total effect strength	57.8	75.4

The following ODA script was used to optimize this LRA model.

```
OPEN ex94.dat;
OUTPUT ex94.out;
VARS behavior yhat;
CLASS behavior;
ATTRIBUTE yhat;
LOO;
GO;
```

The optimized LRA model is, If yhat ≤ .13, then predict that behavior = continue smoking; otherwise, predict that behavior = quit smoking (recall that the cutpoint is .5 for any non-optimized FLDA model). ODA-based optimization improved the classification performance of LRA in training analysis: the strong total effect strength of the optimized LRA model in

LOO validity analysis—which was the same as in training analysis (75.4%)—represented a 30.4% improvement versus the total effect strength obtained by the non-optimized LRA model in training analysis (57.8%).

Example 9.5

As an example of ODA-based optimization of an LRA model involving interval attributes and a hold-out validity sample, reconsider Yarnold, Bryant, et al.'s (1996) data on sympathy and empathy for physicians versus undergraduates (Example 9.1). The total sample was first randomly split into a training and a hold-out validity sample. A participants-to-attributes ratio of between 40-to-1 and 50-to-1 is typically recommended for determining minimum training sample size for applications in which models derived using LRA will be applied to hold-out samples (cf. Wright, 1995). Thus, because four attributes were used, we created a training sample consisting of 4×40, or 160 observations, randomly chosen from the total sample. The hold-out sample, therefore, consisted of 209 − 160, or 49 observations. The training (ex95A.dat) and hold-out (ex95B.dat) data files provide the data for each observation on a separate line: The first variable is person, and the following four variables are fan, per, emp, and dis.

Next, LRA was performed for the training sample. The LRA-derived equation for Y-hat = 3.4009 − .4280 (fan) + .6562 (per) − .9828 (emp) − .6721 (dis). The LRA classification model is, If Y-hat ≤ .5, then predict that person = undergraduate; otherwise, predict that person = physician. This model was applied to the data of the training sample, and to the data of the hold-out sample, and classification performance was determined. For the training and hold-out samples, the following classification performance was achieved by the non-optimized LRA model.

Classification Performance Index	Training	Hold-Out
Overall classification accuracy	73.8	63.3
Sensitivity (undergraduate)	59.2	58.3
Sensitivity (physician)	85.4	68.0
Effect strength, sensitivity	44.5	26.3
Predictive value (undergraduate)	76.4	63.6
Predictive value (physician)	72.4	63.0
Effect strength, predictive value	44.7	26.6
Total effect strength	44.6	26.4

Next, *p* was determined for these hold-out results using the following ODA script (the hold-out data were taken from the output of the preceding analysis). The analysis shows that the hold-out classification performance yielded by the non-optimized LRA model was not statistically significant at the generalized criterion: estimated $p < 0.057$.

```
OPEN DATA;
OUTPUT ex95.out;
CATEGORICAL ON;
TABLE 2;
CLASS ROW;
DIRECTIONAL < 1 2 3;
MCARLO ITER 10000;
DATA;
14 10
8 17
END DATA;
GO;
```

Analysis next shifted to the optimized version of this LRA model. It was first necessary to create two new data files—one for the training sample (ex95C.dat), the other for the hold-out sample (ex95D.dat): Each provided information for each observation on two variables—person and yhat. (Recall that when using ODA software the training and hold-out data file(s) must have identical format.) Because the class sample sizes were not the same, weighting by prior odds was specified. LOO validity analysis was requested. The analysis was accomplished using the following ODA script.

```
OPEN ex95c.dat;
OUTPUT ex95c.out;
VARS person yhat;
CLASS person;
ATTRIBUTE yhat;
LOO;
HOLDOUT ex95d.dat;
GO;
```

The optimized LRA model was, If yhat $\leq .0694$, then predict person = undergraduate; otherwise, predict person = physician. The classification performance achieved by the optimized LRA model in training and hold-out analysis is summarized below.

Classification Performance Index	Training	Hold-Out
Overall classification accuracy	75.00	65.31
Sensitivity (undergraduate)	64.79	62.50
Sensitivity (physician)	83.15	68.00
Effect strength, sensitivity	47.93	30.50
Predictive value (undergraduate)	75.41	65.22
Predictive value (physician)	74.75	65.38
Effect strength, predictive value	50.16	30.60
Total effect strength	49.05	30.55

Compared to the non-optimized model, the optimized model resulted in 10.1% improvement in total effect strength in training analysis and 15.9% improvement in total effect strength in hold-out analysis.

Finally, p was estimated for the hold-out results of the optimized LRA model. This was accomplished via the same script used to estimate p for the hold-out results of the non-optimized model, except that the data were "15 9 8 17". For this analysis, the hold-out classification performance of the optimized model was very near to—but was not less than—the generalized criterion for statistical significance: estimated $p < 0.0522$ (based on 50,000 Monte Carlo experiments, estimated $p < 0.0533$).

Optimizing Complex Models

All of the multiattribute models provided in this chapter were "simple effects" models. For example, imagine an application involving two attributes: A1 and A2. A simple effects model would attempt to predict the class variable using only A1 and A2. A fully saturated model, in contrast, would attempt to predict the class variable using A1 and A2 as attributes, but also using their interaction (A1 × A2). Furthermore, it is also possible to use squared terms ($A1^2$, $A2^2$) as attributes. Whereas the use of these additional attributes may increase the PAC of the model, their inclusion does not alter the manner in which the resulting model is optimized via ODA.

Finally, in applications involving more than two class category levels, ODA-based optimization is possible so long as the multicategory multiattribute suboptimal model provides one classification function. For a problem involving C class categories, this would correspond to a single classification function with $C - 1$ cutpoints used to define class categories. In such cases, ODA-based optimization proceeds as was illustrated in this chapter, but using the multicategory techniques discussed in chapter 6.

CHAPTER 10

Multiple Sample Analysis

*I*n multisample applications involving two or more samples, all observations must have data on (preferably) an identical or (minimally) a parallel set of class variables, attributes, and weights. Multiple samples may be drawn from a single population, such as samples of cardiologists drawn from different hospitals, or samples of undergraduates from a psychology department participant pool drawn from different semesters. Alternatively, multiple samples may be drawn from multiple populations, such as samples of male versus female cardiologists (undergraduates) drawn from different hospitals (semesters). ODA methods already discussed may be used to address several important issues involving multiple samples.

First, ODA may be used to assess between-sample differences. That is, by treating sample as a class variable, one can determine whether samples differ on any measured attributes. For example, in clinical research it is often important to know whether treatment and control groups differ with respect to some set of covariates. In this context, ODA can identify covariates which discriminate samples—in terms of both statistical and practical significance.

Second, ODA may be used to assess between-sample similarities. That is, by selecting one sample for use in training analysis and then treating the other data as different hold-out samples, one can determine the extent to which an ODA model developed using the training sample cross-generalizes to the hold-out samples. In applications involving an ODA model that is *fixed*—such as in designs involving a directional hypothesis with a class variable and attribute that have the same number of categories—the choice of which data set to use in training analysis is essentially arbitrary. However, in applications involving an ODA model that has *room to vary*—such as in problems involving continuous attributes—the model cutpoint(s) may differ depending on the sample selected for training analysis.

The latter fact is disquieting: If training model cutpoints may vary depending on the sample selected for analysis, then obviously so too may the resulting model classification performance vary—for both training and hold-out analyses. Imagine a five-sample application: Analyzed separately, the identical ODA model classified observations perfectly for the first four samples, but for the fifth sample a different ODA model classified observations with accuracy comparable to chance. If the fifth sample is selected for use in training analysis, then the entire analysis is compromised. How, then, does one determine the single best ODA model for a multisample application? What is a straightforward, objective means of determining whether the single best ODA model generalizes across the multiple samples? Discussed below, a statistical tool was developed to answer these questions, that is called *Gen*—our laboratory pseudonym for *the generalizability algorithm*.

However, it is important to note at this juncture that research often misanalyzes multisample applications, behaving as though the application actually involved a single sample. For example, a common practice involves pooling data collected from multiple sites to obtain a sufficiently large total sample to ensure reasonable power for statistical hypothesis-testing purposes (e.g., a rehabilitation study involving patients drawn from a dozen facilities or a management study involving employees drawn from a dozen stores). It also is common to incorrectly assume that a single sample obtained from a single site necessarily constitutes a single sample application. In fact, a "single sample" may actually consist of many different samples: male versus female, rich versus poor, married versus single people of different races, creeds, and political affiliations, for example (cf. Yarnold, 1992). This is important because—due to a psychometric anomaly known as Simpson's paradox—*it may be inappropriate to combine the data from multiple sites or samples into a single total data set.*

Pooling Samples and Simpson's Paradox

Simpson's (1951) paradox refers to a problem that, left unchecked, may obstruct the valid analysis of data and interpretation of findings in studies involving multiple samples (Blyth, 1972; Hintzman, 1993; Preuss & Vorkauf, 1997; Yarnold, 1996a). Development of meta-analysis was motivated in part by difficulties faced in attempting to combine multiple samples (Rosenthal, 1984). The paradox may arise in research involving multiple samples (such as occur in multisite clinical trials); single samples that in reality combine the data of two or more distinct groups, such as men and women; or a single observation followed over a period of time within which an intervention has occurred. The nature of the paradox is that separate analyses of data conducted for the individual samples, groups, or time periods may yield results which contradict the analysis of data for the pooled samples, groups, or time periods. Such paradoxical confounding can simultaneously serve to suggest invalid conclusions and to mask actual meaningful differences between samples, groups, or time periods.

Simpson's paradox may occur for applications involving only categorical data: "If two or more contingency tables are collapsed into one, the resulting table may show a relationship between variables different from those shown by any of the original tables" (Hintzman, 1980, p. 398). Martin (1981) provided this explanation: "Simpson's paradox is therefore simply an instance where both of the two factors of interest are correlated with a third factor" (p. 372). A remedy for Simpson's paradox in categorical applications involves standardizing (*homogenizing*) individual contingency tables before pooling them and, if information concerning potential "third factor" variables is available, then conducting log-linear analysis analogous to partial correlation (e.g., Bishop, Fienberg, & Holland, 1975; Flexser & Tulving, 1993; Woolson, 1987). Although promising, these techniques may be problematic for designs involving small numbers of observations, one or more sparse cells, or a high degree of sample size imbalance (e.g., Bishop et al., 1975; Dowdney, Rogers, & Dunn, 1993; Hagenaars, 1990).

Simpson's paradox may also occur for applications involving only ordered data (Rosenthal, 1984; Yarnold, 1996a). For example, imagine two hypothetical samples, A and B: Each sample has two observations, and each observation has scores on two attributes, X and Y. For sample A, the first observation's scores are $X = 2$, $Y = 0$, and the second observation's scores are $X = 0$, $Y = 2$. Note that X and Y correlate at $r = -1$ for sample A. For sample B, the first observation's scores are $X = 8$, $Y = 6$; the second observation's scores are $X = 6$, $Y = 8$; and X and Y correlate at $r = -1$. Thus, for both samples considered independently, X and Y are perfectly negatively correlated. Simpson's paradox for this example is, When the data of samples A and B are pooled into a combined (or total) sample consisting of all four observations, X and Y correlate at $r = +.8$.

Analogous to categorical designs, the remedy for Simpson's paradox for ordered data involves normative standardization (see Standardizing Transformations section, chapter 3) of X and Y separately by sample. In the present example, standardizing data in sample A yields $z_X = .71$, $z_Y = -.71$, for the first observation, and $z_X = -.71$, $z_Y = .71$, for the second observation: z_X and z_Y correlate at $r = -1$. Similarly, standardizing data in sample B yields $z_X = .71$, $z_Y = -.71$, for the first observation, and $z_X = -.71$, $z_Y = .71$, for the second observation: z_X and z_Y correlate at $r = -1$. When the standardized data from samples A and B are pooled, there are two observations with scores $z_X = .71$, $z_Y = -.71$, and two observations with scores $z_X = -.71$, $z_Y = .71$: z_X and z_Y correlate at $r = -1$.

This remedy for Simpson's paradox is necessary only if the samples differ with respect to X or Y (otherwise, raw data may be used) and it is successful in ameliorating the paradox only if the relationship between X and Y is consistent across the multiple samples (Yarnold, 1996a). For example, imagine that X and Y were perfectly positively correlated in sample C, and perfectly negatively correlated in sample D. If data were normatively standardized separately by sample and then combined, X and Y would have a correlation of zero (assuming equal sample sizes). Thus, in order to meaningfully combine data from multiple samples, it is *not* sufficient to normatively standardize the data separately by sample—it is also necessary to verify that the relationship between X and Y is homogeneous across samples (cf. McClish, 1992; Midgette,

Stukel, & Littenberg, 1993; Rosenthal, 1984). *Do not pool data if the relation between* X *and* Y *is inconsistent across samples*. Note that this procedure has an analog in the analysis of categorical data: the "iterative proportional fitting" method involves standardizing marginals before combining multiple categorical data samples (e.g., Bishop et al., 1975; Mosteller, 1968).

Paradoxical confounding may also occur for designs involving ordered and categorical data and, as in purely categorical or ordered applications, the remedy involves separate standardization by sample (if necessary), assuming that the relationship between categorical and ordered attribute generalizes across samples (cf. McClish, 1992; Midgette et al., 1993). When it is theoretically possible, the generalizability algorithm enables circumvention of Simpson's paradox regardless of the metrics underlying the class variable and attribute.

The ODA Generalizability Algorithm

In applications where the model cutpoint has room to vary, the generalizability algorithm, *Gen*, identifies the single ODA model that—when simultaneously applied to multiple samples—maximizes the *minimum* overall (non)weighted percentage accuracy in classification (PAC) achieved by the model across the samples. As used here, (non)weighted means "nonweighted or weighted." The ODA model identified in a Gen analysis is referred to as a *GenODA* model.

Imagine a two-sample application having equal class sample sizes and involving a continuous attribute. Further imagine that an ODA model was obtained for the combined (pooled) data that maximized overall (non)weighted PAC regardless of the PAC achieved for either original sample considered separately: For the sake of argument, imagine that the ODA model obtained 100% PAC in one sample, and 50% PAC in the other sample. In contrast, imagine that a GenODA model was obtained that maximized the PAC achieved for the sample that was classified least successfully: Imagine that the GenODA model obtained 70% PAC in one sample, and 60% PAC in the other sample. Although the ODA model had greater overall PAC than the GenODA model (75% versus 65%, respectively), the latter model is optimal under the Gen criterion because the minimum PAC it achieved in the worst-classified sample (60%) was maximized (e.g., versus the minimum PAC of 50% achieved by the ODA model).

Weighted Gen models are also possible. For example, combining Gen and prior odds weighting results in a GenODA model that maximizes the minimum mean sensitivity achieved for any sample by the model when it is simultaneously applied across samples. Combining Gen and return weights results in a GenODA model that maximizes the minimum return-weighted overall PAC achieved by the model for any sample when it is simultaneously applied across samples. And combining Gen with prior odds and return-weighting results in an ODA model that maximizes the minimum mean return-weighted sensitivity achieved by the model for any sample when it is simultaneously applied across samples.

GenODA models provide a solution for problems described earlier involving multisample analysis. First, GenODA determines the single best ODA model for a multisample application in the sense that the resulting model ensures that the worst classification performance achieved for any sample is maximized. Second, GenODA provides a straightforward, objective means of determining whether the single best ODA model generalizes across multiple samples. That is, if the effect strength achieved for the worst-classified sample is lower than the a priori user-defined criterion for successful generalizability, and/or if the corresponding p achieved for the worst-classified sample is greater than the a priori user-defined criterion for statistical significance, then it is concluded that the GenODA model did not achieve satisfactory performance for all samples. It may therefore be concluded that no single ODA model can achieve satisfactory classification performance for all of the samples: In this circumstance, data cannot be meaningfully pooled for all of the samples. In contrast, If the GenODA model exceeds the minimum a priori standards for acceptable effect strength and/or p for the worst-classified sample, then it is concluded that the GenODA model achieved satisfactory performance for all samples.

Finally, GenODA constitutes a test for the absence of paradoxical confounding. That is, the finding that a GenODA model achieves statistically and/or practically significant classification accuracy, even in the worst case, supports the alternative hypothesis that the model has validity when applied across the different samples (the null hypothesis is that this is not true). In contrast, the finding that a GenODA model is not satisfactory in the worst case implies that the model fails to apply across all of the samples: This suggests the possibility of paradoxical confounding if data are pooled and raises the issue of whether any ODA model applies for any of the samples.

Before proceeding to illustrate how to evaluate the generalizability of a GenODA model when it is simultaneously applied across multiple samples, we first describe two heuristics that may be used in addition to those described in chapter 3 as a means of selecting from among multiple optimal Gen solutions, were they to occur for a given application.

First, the *Gen PAC of Sample s* selection heuristic retains the GenODA model that achieves maximum (non)weighted PAC for the particular sample (sample s) that is specified by the user. For example, imagine a two-sample (1 and 2) application for which the accurate classification of observations from sample 1 is of paramount priority. Further imagine that two optimal models emerged (models Y and Z), and that both achieved overall PAC = 75%. Finally imagine that, for sample 1, model Y achieved 50% PAC and model Z achieved 60% PAC. Model Z would be selected by the Gen PAC of sample 1 heuristic, because model Z had greatest PAC for sample 1.

Second, the *Gen maximum mean PAC* heuristic, also known as the *prior odds of samples* heuristic, retains the GenODA model that achieves maximum (non)weighted mean overall PAC across samples. For example, imagine a two-sample application with sample 1 having 10 observations, and with sample 2 having 20 observations. Further imagine that two optimal models (Y and Z) emerged—each achieving an optimal value of 20 correct classifications (overall PAC = 67%). Imagine that model Y misclassified all sample 1 observations, and correctly classified all sample 2 observations: sample 1 PAC = 0%; sample 2 PAC = 100%; mean PAC across samples = 50%. Finally, imagine that model Z correctly classified all sample 1 observations, and correctly classified half of the sample 2 observations: sample 1 PAC = 100%; sample 2 PAC = 50%; mean PAC across samples = 75%. Because the mean PAC across samples is greatest for model Z, it would be selected by the Gen maximum mean PAC heuristic.

Evaluating Model Generalizability Across Samples

In some applications a GenODA model is structurally predetermined by the combination of data metric and directional hypothesis: Such models are referred to as being *fixed*. For example, consider (a) a five-category class variable, father's occupational class; (b) a five-category polychotomous attribute, son's occupational class (father's and son's occupational class codes are identical); and (c) the directional alternative hypothesis that fathers and sons have identical occupational class codes. This situation decrees the directional GenODA model for predicting father's occupational class: If son's occupational class code = 1, then predict that father's occupational class code = 1; if son's occupational class code = 2, then predict that father's occupational class code = 2; and so forth. The cutpoints for this GenODA model are fixed: Irrespective of the number of samples involved, the fixed GenODA model will remain structurally invariant.

For a fixed model, a GenODA analysis performed via ODA software provides a separate test of the directional hypothesis for each separate sample, as well as for the pooled (combined) data. To evaluate the generalizability of a fixed-model GenODA model across samples, first—prior to conducting analysis—decide on the minimally acceptable level of effect strength and/or p that must be achieved for each separate sample, and for the pooled data. Then, after obtaining the GenODA model, evaluate the minimal acceptability criterion separately for each sample and for the pooled data. If all samples (including pooled) meet this criterion, then conclude that the GenODA model generalized across samples. Results for the pooled sample represent an omnibus test of the directional hypothesis across samples. However, if at least one sample fails the minimal acceptability criterion, then conclude that the GenODA model does not generalize across samples, and do not interpret the results for the pooled sample (options that are available once this latter situation is encountered are discussed shortly).

Example 10.1

As an example of evaluating the generalizability of a fixed GenODA model across multiple samples, we return to the data of Example 8.1, in which the generalizability of the association between father's and son's occupational status (coded using five categorical levels consistent between father and son and across country) is assessed with data from England and Denmark. A cross-classification table was created separately for each country, with rows reflecting father's occupational status and columns reflecting son's occupational status. Data were originally entered using two files, because one country was being treated as a hold-out sample for the other country. In the present example, however, data for both countries were entered into a single file using free tabular input: English data were entered first, followed by Danish data. The directional alternative hypothesis tested in Example 8.1—that fathers and sons have identical occupational class codes—was also tested here. Weighting by prior odds was used because the different occupational codes are associated with different numbers of observations. LOO validity analysis is always superfluous for fixed ODA or GenODA models. Finally, the minimal acceptability criteria for all three (British, Danish, pooled) analyses include (a) to attain total effects of moderate strength or better and (b) because of the large samples, to satisfy the sequential Sidak criterion for experimentwise $p < 0.05$. Following is the ODA script used for this analysis.

```
OPEN ex101.dat;
OUTPUT ex101.out;
CATEGORICAL ON;
TABLE 5;
CLASS ROW;
GEN TABLE 2;
DIRECTIONAL < 1 2 3 4 5;
MCARLO ITER 10000 TARGET .05 SIDAK 3;
GO;
```

The total effect strength for the directional model was 23.50% for the British sample (coded as 1) and 23.62% for the Danish sample (coded as 2), and both were statistically significant with experimentwise $p < 0.05$. Both models had estimated $p < 0.0001$, with 99.99% confidence for $p < 0.01696$ (the Sidak criterion). Finally, for pooled data the GenODA model was statistically significant (estimated $p < 0.0001$), with total effect strength = 24.47%.

Thus, all three analyses exceeded the statistical significance criterion of an experimentwise $p < .05$, but came marginally short of meeting the practical significance criterion of a total effect strength of 25% or higher. These results are thus only partially consistent with the directional hypothesis concerning the direct relationship between fathers' and sons' occupational class. Whereas this hypothesis found statistical support for the separate samples and for the pooled data, the practical effect strengths in these analyses were marginally lower than desired.

In contrast to the fixed GenODA model, *room-to-vary* GenODA models are not structurally predetermined. For example, imagine (a) a binary class variable, whether a patient is satisfied or dissatisfied with overall care they received (coded as 1 versus 0, respectively); (b) an interval attribute, score on a satisfaction with physician (SWP) scale (ranging between 5 and 30); and (c) the directional alternative hypothesis that satisfied patients will be more satisfied with their physician (i.e., will have a greater SWP score) than dissatisfied patients. This situation decrees a directional GenODA model for predicting patient overall satisfaction: If SWP score ≤ cutpoint, then predict dissatisfied patient; otherwise, predict satisfied patient. The cutpoint for this GenODA model is not fixed, but instead has room to vary: The specific value of the cutpoint that maximizes model performance for any particular division of medicine (e.g., cardiology, internal medicine, etc.) will likely be different for different divisions (samples), and different cutpoints may be associated with differential classification performance. Depending on the number of samples involved, the room-to-vary GenODA model will *not* remain structurally invariant. The GenODA model maximizes the (non)weighted classification performance for the sample having the least successful results.

As for a fixed model, a room-to-vary GenODA analysis performed via ODA software provides a separate test of the directional hypothesis for each separate sample, as well as for the pooled (combined) data. In order to evaluate the generalizability of a room-to-vary GenODA model across samples, first—prior to conducting analysis—decide on the minimally acceptable level of effect strength and/or p that must be achieved for each separate sample, and for the pooled data. If the application involves a nondirectional hypothesis and either a polychotomous or ordinal attribute having the same number of levels as there are categories of the class variable, then proceed with the analysis as described for fixed GenODA models, and perform LOO validity analysis.

However, for room-to-vary applications involving continuous (interval or ratio) attributes, there are two levels at which the test of the generalizability of the directional hypothesis may be assessed. First, *strict generalizability* occurs if a GenODA model fits *raw* data for all samples. In this instance, this would indicate that a specific numerical value for raw SWP score successfully discriminates satisfied versus dissatisfied patients independently of division. Second, *relative generalizability* occurs if a GenODA model fits *normatively standardized* data (standardization performed separately by sample) for all samples: In this instance, this would indicate that a specific numerical value for z_{SWP} successfully discriminates satisfied versus dissatisfied patients independently of division (z_{SWP} may be converted back to raw units separately by division).

To conduct generalizability analysis for an application involving a room-to-vary GenODA model and a continuous attribute (whether or not the hypothesis is directional), first assess whether evaluation of the strict generalizability hypothesis is appropriate. That is, use ODA to evaluate the nondirectional alternative hypothesis that the different samples (treated as a class variable) can be discriminated on the basis of the attribute. If the samples may be discriminated, then it is necessary to normatively standardize the attribute separately by sample in order to avoid Simpson's paradox and assess relative generalizability. However, if the samples may not be discriminated, then raw data may be used to assess the strict generalizability hypothesis. Next, continue the analysis as was described for fixed GenODA models and perform a LOO validity analysis.

Example 10.2

As an example of evaluating the generalizability of a directional room-to-vary GenODA model across multiple samples, consider Thompson et al.'s (1996) data concerning the association between patient overall satisfaction and SWP score for five different divisions of medicine: allergy, arthritis, cardiology, dermatology, and gastroenterology (respectively coded as 1 to 5). The directional alternative hypothesis is that, relative to dissatisfied patients, patients who are satisfied with their overall care (binary class variable) will be more satisfied with their physician—that is, will have greater SWP scores (continuous attribute). Weighting by prior odds is used because there are different numbers of (dis)satisfied patients, and LOO validity analysis is conducted to assess model stability. Finally, because prior research found that satisfaction with physician is strongly related to overall satisfaction, the minimal acceptability criteria for all analyses include (a) strong total effects (i.e., 50% or greater) and (b) satisfy the sequential Sidak criterion for experimentwise $p < 0.05$ (the appropriate sequential Sidak target is specified in MCARLO when mop-up analyses are conducted—that is, after the overall number of effects to be reported in the manuscript has been determined).

The first step of the analysis involves determining whether it is appropriate to evaluate the strict generalizability hypothesis. Accordingly, the following ODA script was used to assess whether samples could be discriminated on the basis of the attribute. As a pragmatic issue, nondirectional models involving an ordered attribute having five or more class categories, a relatively large sample (several hundred observations) and a nondirectional hypothesis are computational nightmares. Thus, when confronted with such an application, we initially run an ODA model without simulation or LOO analysis: We simply wish to determine the structure of the ODA model that maximizes (non)weighted PAC.

```
OPEN ex102a.dat;
OUTPUT ex102a.out;
VARS overall division swp;
CLASS division;
ATTRIBUTE swp;
GO;
```

The ODA model was obtained from the output and entered in a DIRECTIONAL command. The following ODA script was appended to the previous program.

```
DIRECTIONAL < 4 5 2 1 3;
LOO;
MCARLO ITER 1000;
GO;
```

The use of a directional hypothesis always reduces computation time because such a constraint limits the number of directions that the model must explore in order to identify the optimal solution. In general, the larger the number of constraints in one's model, all other things being equal, the faster the solution time. Use of a directional hypothesis vastly speeds computation time in the present example (even so, this problem takes substantially greater time to solve than prior examples, as evidenced by the use of only a thousand Monte Carlo iterations). Here, estimated p will appear smaller than it actually is, because the appropriate analysis is nondirectional. Also, estimated LOO classification performance may appear greater than it actually is, because in a nondirectional hypothesis the model is free to vary, introducing instability. This directional analysis thus provides an upper-bound estimate of the discriminability of the multisample variable. In this instance, the resulting classification performance was weak in practical terms (total effect strength = 15.44%) and it degraded in LOO analysis (total effect strength = 13.00%). Because this effect is very weak and the likelihood of paradoxical confounding is therefore low, we tested the strict generalizability hypothesis using raw data. (Not done here. For readers who wish to compare the findings of this analysis with an assessment of relative generalizability, separately normatively standardized SWP data are provided in ex102b.dat.) The following ODA script was appended to the end of the above program.

```
CLASS overall;
GEN division;
DIRECTIONAL < 0 1;
LOO;
MCARLO ITER 10000;
GO;
```

The GenODA model was If SWP ≤ 29.5, then predict that the patient is dissatisfied; otherwise, if SWP > 29.5, then predict that the patient is satisfied. The following classification performance was achieved:

| | Total Effect Strength | |
Division	Training	LOO
1	32.6	32.6
2	51.0	51.0
3	4.7	−10.1
4	33.9	33.9
5	52.4	52.4

Only divisions 2 and 5 passed the effect strength acceptability criteria. Classification performance was stable for all divisions except 3: Clearly, division 3 is unusual. Accordingly, a second GenODA analysis was conducted, eliminating division 3, by appending the following script at the end of the previous program.

```
EXCLUDE division = 3;
GO;
```

The GenODA model was, If SWP ≤ 21.5, then predict that the patient is dissatisfied; otherwise, if SWP > 21.5, then predict that the patient is satisfied (recall that with division 3 in the model, the cutpoint was 29.5). The following classification performance was achieved:

| | | Total Effect Strength | |
Division	Training $p <$	Training	LOO
1	.0001	62.5	58.8
2	.0001	63.9	63.9
4	.0001	56.4	56.4
5	.0001	59.8	59.8

Every effect strength index is stronger in this analysis than in the preceding analysis, and all analyses have effect strength and associated p that exceed the minimal acceptability standards (classification performance degraded marginally for division 1). Thus, classification performance achieved on the basis of the directional hypothesis that satisfied patients have higher SWP scores than dissatisfied patients was relatively strong in practical terms, was statistically significant at experimentwise $p < .05$, and cross-generalized across four of the five divisions.

It is unclear whether a test for relative generalizability would have fit division 3 data. This issue may be answered by determining whether *any* ODA model will fit division 3 data. That is, if an ODA model specifically optimized for division 3 fails to fit, then division 3 cannot be fit by any (Gen)ODA model. This analysis was accomplished by appending the following ODA script at the end of the previous program.

```
GEN OFF;
INCLUDE division = 3;
GO;
```

The resulting ODA model was structurally identical to the GenODA model from the initial (five sample) analysis: Involving a cutpoint of 29.5 ($p < 0.53$), this model dominated the original GenODA analysis. Of course, the classification performance was therefore also the same as was identified in the original GenODA analysis. Eliminating the directional hypothesis did not help matters. Apparently, overall patient satisfaction for cardiology patients is not predictable on the basis of their SWP scores. Perhaps this is attributable in part to range restriction in their SWP scores, because the mean score for division 3 is very near ceiling: mean SWP score = 29.0, corresponding to 96.7% of theoretical maximum for this scale.

Analyzing Randomized Block Designs

The massive literature addressing analysis of variance (ANOVA) is too large to be reviewed here (Box & Draper, 1987; Brownlee, 1960; Cochran & Cox, 1957; Fisher, 1953; Fleiss, 1986; Hays, 1973; Hicks, 1973; Kempthorne, 1952; Keppel, 1982; Kirk, 1982; Kramer, 1972; Maxwell & Delaney, 1990; Mosteller & Tukey, 1977; Scheffe, 1959; Stevens, 1992; Winer, 1971; Woolson, 1987). However, the working hypothesis in our laboratory is that an ODA analog or alternative exists for every ANOVA model. As shown in chapters 5, 6, and elsewhere in this book, ODA can be used to analyze data configurations that reflect a completely randomized design: typically analyzed using *t* test and one-way ANOVA. However, ODA may also be used to analyze data reflecting experimental designs that are traditionally analyzed using more complex ANOVA models.

The randomized block design (RBD) is a popular ANOVA model that uses a blocking procedure in an attempt to reduce error variance and thereby increase power (Kirk, 1982). For example, in medical research comparing the effectiveness of different treatments, subject variables such as age, race, or gender may serve to increase error variance and thus reduce or obscure the effect of the treatments. Similarly, in military research comparing the effectiveness of different anti-aircraft warheads, environmental variables such as wind, temperature, or atmospheric radiation may increase error variance and reduce the effect attributable to the warheads. Variation attributable to subject, environmental, or other conceptually similar "nuisance" variables is referred to as *nuisance variation*. In the RBD method, the nuisance variable is included as one of the factors in the experiment, and a blocking procedure is used in order to partition the nuisance variation out of the ANOVA estimates for both the treatment and the error terms: The treatment effect is treated as fixed, and the block effect as random. A conceptually similar approach is used in meta-analytic research in an attempt to discover variables that explain discrepancies or inconsistencies in the results of different studies that investigate the same phenomenon (Rosenthal, 1984).

In the typical RBD data configuration, rows correspond to the blocking (nuisance) variable and columns correspond to the treatments: A minimum of two blocking variables and two treatments are required. When constructing blocks, it is essential that observations within a block are more homogeneous than are observations between different blocks: If this condition is not satisfied, then the blocking variable will not fulfill its intended purpose of explaining nuisance variation, reducing the error term, and increasing power. Because of this requirement, the repeated measures approach—in which every observation is exposed to each treatment—is especially popular, because in this approach observations serve as their own controls and thus maximize within-block homogeneity (a repeated-measures RBD is also referred to as a *subjects-by-treatments design*). Finally, when there are multiple observations per block, there should be random assignment of the order of presentation of the treatment levels to the observations within each block, and when there is only one observation per block the order of presentation of the treatment levels should be randomized independently for each observation.

The RBD ANOVA provides two *F* statistics: one for treatments (the null hypothesis is that the mean value of the attribute is the same across treatments) and one for blocks (the null hypothesis is that the mean value of the attribute is the same across blocks). Blocks in an RBD are selected so as to be heterogeneous, such that it is expected that there will be a statistically significant block effect. Thus, the block effect is typically of little theoretical interest outside of its intended role of reducing the error term. It should be noted that this RBD is additive, implying that treatment and blocking variables do not interact. If nonadditivity is assumed, then the interaction between treatment and blocking variables must be included in the ANOVA. This type of RBD is not considered here because it is a MultiODA problem. If there are more than two treatment levels and the null hypothesis is rejected, then additional contrasts are required to ascertain the precise nature of the interclass differences. Finally, it is important to assess the extent to which the assumptions underlying ANOVA have been violated: Such violations can reduce power or, worse, render *p* invalid (Bradley, 1968). The impressive number and complexity of assumptions required of data to validly use the RBD renders unlikely the possibility that they will be completely satisfied in most real-world applications (cf. Kirk, 1982).

GenODA may be used to analyze data from an additive RBD. To accomplish this, treatment is configured as the class variable and the blocking variable as the Gen (multisample) variable. For ODA used in this context the null hypothesis is that the attribute cannot be used to discriminate the classes in a manner consistent across the different levels of the blocking variable. If this null hypothesis is rejected, the implication is that the attribute can be used to reliably discriminate the classes in a consistent manner across the blocking variable:

a compelling argument in support of the generalizability of the effectiveness of the treatment across blocking strata. If, however, this null hypothesis is not rejected, it is unclear whether this is because the attribute is simply not predictive of class membership or because the attribute is not predictive of class in a consistent manner across the different blocks. Thus, failing to reject the null hypothesis in an ODA analysis of an RBD is not a particularly compelling result (of course, this is generally true of any statistical procedure). In such a case, if the investigator wishes to identify subgroups with respect to the blocking strata that do share a common relationship between the attribute and the class variable within subgroup, then the methodology described in Example 10.2 could be used. Finally, in contrast to the assumption-laden ANOVA approach, ODA requires no distributional assumptions.

Example 10.3

As an example of the use of GenODA to analyze data from a nondirectional RBD, consider Arthanari and Dodge's (1981) data, involving modeling the yield (kilograms per 100 square meters) of three varieties of wheat (Karoon = 1, Shoeleh = 2, Mexican = 3) as a function of two types of fertilizer (dummy-coded as 1 and 2). The issue is whether or not the different types of fertilizer result in different yields of wheat. However, because three different types of wheat (samples) were examined, and because type of wheat may be related to yield, the possibility of Simpson's paradox exists. Because of the small sample size and low power and the exploratory nature of the research, the acceptability criterion for statistical significance was generalized $p < .10$. Also because of the small sample, the acceptability criterion for practical significance was that the total effect strength should be relatively strong (50% or greater). Prior odds weighting was used because fertilizer class sample sizes were imbalanced. LOO analysis was not requested because there must be at least two observations per class category per sample.

First, ODA was performed to determine if wheat variety can be discriminated on the basis of yield—which would indicate the possibility of paradoxical confounding. This analysis was accomplished using the following ODA script.

```
OPEN ex103a.dat;
OUTPUT ex103a.out;
VARS fertilizr variety yield;
CLASS variety;
ATTRIBUTE yield;
LOO;
MCARLO ITER 10000;
GO;
```

The ODA model, which achieved perfect overall PAC, was, If yield ≤ 13, then predict that variety = 2 (Shoeleh); otherwise, if 13 < yield ≤ 29.5, then predict that variety = 1 (Karoon); otherwise, if yield ≤ 29.5, then predict that variety = 3 (Mexican). This model was statistically significant: estimated $p < 0.0004$. The finding that the three wheat varieties can be perfectly discriminated on the basis of yield suggests possible paradoxical confounding.

Thus, yield data were normatively standardized (zyield) separately by variety (had these varieties of wheat instead not been discriminable on the basis of yield, no standardizing transformation would have been needed and raw data could have been used). GenODA was then used to determine whether type of fertilizer (three-category class variable) can be discriminated on the basis of standardized yield (continuous attribute) across variety of wheat (sample). This analysis was accomplished via the following script.

```
OPEN ex103b.dat;
OUTPUT ex103b.out;
VARS fertlizr variety zyield;
CLASS fertlizr;
ATTRIBUTE zyield;
GEN variety;
MCARLO ITER 10000;
GO;
```

The GenODA model was, If zyield ≤ 0.3, then predict fertilizer 2; otherwise, predict fertilizer 1. This model indicates that, irrespective of type of wheat, use of fertilizer 1 resulted in greater wheat yield than did use of fertilizer 2. This model met the statistical significance acceptability criterion for the pooled sample: estimated $p < 0.011$. Although this criterion was not satisfied when this analysis was conducted separately by variety, it could not be met *in principle* because the sample sizes were too small to yield effects having $p < 0.10$. Thus, here this criterion applies only to the pooled sample analysis, due to tiny statistical power. The model met the practical significance acceptability criterion for Shoeleh wheat (total effect strength = 50%), and exceeded this criterion for Karoon and Mexican wheat (total effect strengths = 100%).

Optimizing Multiple Suboptimal Multiattribute Models

Finally, Gen may also be used to optimize suboptimal multiattribute classification models in multisample applications. Two different GenODA-based optimization methods are considered next. First considered are applications in which a single sample is used to develop the multiattribute model, and several hold-out samples are used to assess its validity. We next consider applications for which separate sample-specific multiattribute models are developed.

Example 10.4

As an example of GenODA-based optimization of a single training model and one or more hold-out samples, consider Yarnold's (1987) research discriminating Type A (coded as 1) versus Type B (coded as 0) undergraduates (the binary class variable) on the basis of two ordered personality dimensions—instrumentality and expressiveness (the continuous attributes). For illustrative purposes, data for the total sample of 600 undergraduates were randomly assigned to one of five subsamples (dummy-coded as 1 to 5). Next, sample 2 was randomly selected to serve as the training sample, and logistic regression analysis (LRA) was conducted using these data to develop a training classification model. The classification performance of the LRA model is summarized below:

Classification Performance Index	Sample 2	Samples 1 and 3–5	All 5 Samples
Overall classification accuracy	83.6	74.9	76.5
Sensitivity (Type A)	80.4	70.0	71.9
Sensitivity (Type B)	86.4	79.2	80.6
Effect strength, sensitivity	66.8	49.2	52.4
Predictive value (Type A)	83.7	74.9	76.5
Predictive value (Type B)	83.6	74.9	76.5
Effect strength, predictive value	67.2	49.8	53.0
Total effect strength	67.0	49.5	52.7

Summarized separately by sample, the training classification performance achieved by the LRA model was

| Classification Performance Index | Sample | | | | |
	1	2	3	4	5
Overall classification accuracy	78.1	83.6	78.3	71.8	71.1
Sensitivity (Type A)	79.6	80.4	71.2	63.5	66.1
Sensitivity (Type B)	77.0	86.4	87.5	77.8	75.9
Effect strength, sensitivity	56.6	66.8	58.8	41.2	42.0
Predictive value (Type A)	69.6	83.7	88.1	67.4	72.5
Predictive value (Type B)	85.1	83.6	70.0	74.7	69.8
Effect strength, predictive value	54.8	69.2	58.2	42.0	42.4
Total effect strength	55.7	68.0	58.5	41.6	42.2

Next, priors-weighted GenODA-based optimization was conducted using the following ODA script.

```
OPEN ex104.dat;
OUTPUT ex104.out;
VARS sample abtype yhat;
CLASS abtype;
ATTR yhat;
GEN sample;
LOO;
GO;
```

The resulting GenODA model was, If yhat ≤ .54674, then predict Type B; otherwise, predict Type A. The training classification performance achieved by this model for sample 2, for all of the other samples, and for all five samples is summarized below.

Classification Performance Index	Sample 2	Samples 1 and 3–5	All 5 Samples
Overall classification accuracy	86.4	75.3	77.3
Sensitivity (Type A)	80.4	66.1	68.7
Sensitivity (Type B)	91.5	83.5	85.0
Effect strength, sensitivity	71.9	49.6	53.6
Predictive value (Type A)	89.1	78.0	80.1
Predictive value (Type B)	84.4	73.6	75.5
Effect strength, predictive value	73.5	53.4	55.6
Total effect strength	72.7	51.5	54.6

The training overall classification accuracy and all three effect strength indices are greater for the optimized LRA model than for the non-optimized LRA model—for sample 2, for all of the samples except for sample 2, and for each sample except 3 and 4 considered separately. This remained true given marginally degraded classification performance of the optimized model in LOO analysis: Over all five samples, overall classification accuracy = 76.67%; effect strength for sensitivity = 52.34%; effect strength for predictive value = 54.08%; and total effect strength = 53.21%. Training classification performance achieved by the optimized model by sample was

| | Sample | | | | |
Classification Performance Index	1	2	3	4	5
Overall classification accuracy	80.5	86.4	74.4	75.8	70.2
Sensitivity (Type A)	77.6	80.4	64.4	61.5	62.5
Sensitivity (Type B)	82.4	91.5	87.5	86.1	77.6
Effect strength, sensitivity	60.0	71.9	51.9	47.6	40.1
Predictive value (Type A)	74.5	89.1	87.0	76.2	72.9
Predictive value (Type B)	84.7	84.4	65.3	75.6	68.2
Effect strength, predictive value	59.2	73.5	52.4	51.8	41.1
Total effect strength	59.6	72.7	52.1	49.7	40.6

Example 10.5

As an example of GenODA-based optimization of multiple training models, reconsider the data of the preceding example, in which an LRA model was obtained for a training sample, and then used to compute Y-hats for observations from four hold-out samples. For the present example, in contrast, a sample-specific LRA model was obtained separately for each sample. Then, the LRA model for the analysis of data from sample 1 was used to obtain Y-hats for participants in sample 1; the LRA model for the analysis of data from sample 2 was used to obtain Y-hats for participants from sample 2, and so forth. The training classification performance of the LRA models is summarized separately by sample in the following example.

| | Sample | | | | |
Classification Performance Index	1	2	3	4	5
Overall classification accuracy	82.9	83.6	77.5	73.4	73.7
Sensitivity (Type A)	77.6	80.4	82.2	63.5	71.4
Sensitivity (Type B)	86.5	86.4	71.4	80.6	75.9
Effect strength, sensitivity	64.0	66.8	53.6	44.0	47.2
Predictive Value (Type A)	79.2	83.7	79.0	70.2	74.1
Predictive Value (Type B)	85.3	83.6	75.5	75.3	73.3
Effect strength, predictive value	64.6	67.2	54.4	45.6	47.4
Total effect strength	64.3	67.0	54.0	44.8	47.3

Next, GenODA-based optimization was conducted using the ODA script for Example 10.4, except that "ex105.dat" was substituted in the OPEN command, and "ex105.out" was substituted in the OUTPUT command. The GenODA model was, If yhat ≤ .54001, then predict Type B; otherwise, predict Type A. The classification performance achieved using this model is summarized separately by sample in the following example.

| | | | Sample | | |
Classification Performance Index	1	2	3	4	5
Overall classification accuracy	82.1	86.4	78.3	75.8	74.6
Sensitivity (Type A)	75.5	80.4	80.8	61.5	69.6
Sensitivity (Type B)	86.5	91.5	75.0	86.1	79.3
Effect strength, sensitivity	62.0	71.9	55.8	47.6	49.0
Predictive Value (Type A)	78.7	89.1	80.8	76.2	76.5
Predictive Value (Type B)	84.2	84.4	75.0	75.6	73.0
Effect strength, predictive value	62.9	73.5	55.8	51.8	49.5
Total effect strength	62.4	72.7	55.8	49.7	49.2

Except for sample 1, the optimized LRA model was superior to non-optimized LRA models in classification performance.

CHAPTER 11

Sequential Analyses

*E*xcept for the thirty hypothetical examples presented in chapter 1, and the discussion of temporal reliability, applications discussed thus far have been *static,* or motionless, in the sense that data were collected at a single point in time. In contrast, in this chapter we consider several different *dynamic* methods that are designed to model motion across time: Markov models, turnover tables, autocorrelation, and single-subject (case) analysis. For techniques discussed here to be appropriate, one's design must involve sequential data—that is, there must be multiple recordings (at least two) of the same attribute for every observation.

Identifying Structure in Markov Transition Tables

Markov models are widely used to study sequential data in fields such as biology, computer science, economics, education, engineering, medicine, oceanography, physics, political science, psychology, and sociology (e.g., Beck & Pauker, 1983; Billingsley, 1961; Coombs, Dawes, & Tversky, 1970; Disney, 1971; Hanneman, 1988; Kemeny & Snell, 1976; Kruskal, 1983; Parzen, 1962; Rau, 1970; Raush, 1965). To use this approach, one must first exhaustively define all of the unique states—that is, conceptually different events—that it is possible to observe in one's specific application: Every unique state is assigned a unique identification code. The initial data for a Markov model constitute a sequence—a continuous consecutive stream—of coded events. In this sequential data stream, every time that any discrete event occurs, it is indicated in the relative order in which it occurred using the appropriate identification code.

In order to illustrate these concepts, imagine that one were interested in investigating the sequential behavior—that is, the daily change in closing price—of a given stock between some initial first date (d1) and some final end date (de). The closing price of the stock on the day before d1—that is, on day d0—serves as the initial reference or starting point. At the end of d1, the price of the stock at d0 ($d0) is subtracted from the closing price of the stock on d1 ($d1). That is, change in price for d1 = Δ $d1 = $d1 − $d0. When Δ $d1 > 0, the price of the stock relative to the day before increased (up); when Δ $d1 < 0, the price of the stock relative to the day before decreased (down); and when Δ $d1 = 0, the price of the stock relative to the day before did not change (same). This process is continued until every possible sequential pairwise difference score is computed. In the present example there are ten such difference scores, with the last one being Δ $d10. Continuing the example, imagine that $d0 = $10; that daily closing prices were recorded for ten days (de = d10); and that the following closing prices were observed for d1 to d10, respectively: $11, $12, $12, $13, $14, $9, $7, $7, $9, $10. Here, the ten consecutive difference scores and their corresponding states (in parentheses) are

Δ $d1 = $d1 − $d0 = $11 − $10 = $1 (up)
Δ $d2 = $d2 − $d1 = $12 − $11 = $1 (up)
Δ $d3 = $d3 − $d2 = $12 − $12 = $0 (same)
Δ $d4 = $d4 − $d3 = $13 − $12 = $1 (up)
Δ $d5 = $d5 − $d4 = $14 − $13 = $1 (up)
Δ $d6 = $d6 − $d5 = $9 − $14 = −$5 (down)
Δ $d7 = $d7 − $d6 = $7 − $9 = −$2 (down)
Δ $d8 = $d8 − $d7 = $7 − $7 = $0 (same)
Δ $d9 = $d9 − $d8 = $9 − $7 = $2 (up)
Δ $d10 = $d10 − $d9 = $10 − $9 = $1 (up)

Once data have been coded, the next step involves creating a *transition table*. In this table the number of rows and the number of columns both equal the number of uniquely coded states. Thus, in the present example, rows and columns both have the same three levels—up, same, and down. The rows of the transition table indicate the state at step i in the sequence, and the columns represent the state at step $i + 1$ in the sequence. The transition table is created by separately evaluating every consecutive pair of i and $i + 1$ states and placing a tally in the cell that describes their relation. For example, on d1 (state i) the price of the stock went up, and on d2 (state $i + 1$) the price of the stock also went up. Thus, for this comparison a single tally would be indicated in the cell corresponding to row = up, column = up. By evaluating, tallying, and cumulating all nine consecutive pair-wise comparisons for the present example, the following transition table emerged:

| | Price Change at Day $i+1$ | | |
Price Change at Day i	Up	Same	Down
Up	3	1	1
Same	2	0	0
Down	0	1	1

In conventional Markov analyses, this transition table is next transformed into a *transition matrix* by dividing the number in each cell of the table by the sum of all of the numbers in the table. For example, for the cell (row = up, column = up), the observed frequency for the cell (3) would be divided by the total number of events recorded in the table (9), resulting in a *state transition probability* for this cell of 3/9, or 0.333. Broadly speaking, Markov models attempt to find order—or structure—in transition matrices. *Markov process models* are appropriate when the state transition probabilities *are sequentially dependent* (i.e., systematically vary over the course of the sequence, such as on the basis of elapsed time or number of cycles), and *Markov chain models* are appropriate when the state transition probabilities are *stationary,* or constant, over the course of the sequence. Although a discussion of the details underlying these methods falls outside of the range of this book, suffice it to say that these analyses may become untenably complicated and assumption-laden in applications involving even one state transition probability that is very near or equal to zero. Transition probabilities may be zero for theoretical reasons that render the cell logically impossible (a *structural zero*), or because that is simply how the data happened to occur (an *empirical zero*). Troublesome conventional analyses are generally the rule when the transition matrix is sparse—that is, when there are many small probabilities. Fortunately, ODA may be used to ascertain the structure underlying transition tables.

Example 11.1

As an example of the use of ODA to identify structure underlying transition tables, consider Driese and Dott's (1984) data concerning 22 stratigraphic sections from the Wasatch and Uinta Mountains. They used a modified Markov model to determine whether the order of stacking of different types of rock within carbonate units is random. The data consisted of a transition table involving seven different types of rock (states) and a total of 514 state transitions (shifts from one type of rock to another). The rock types were bioclastic grainstone to packstone (dummy-coded as A), pelletoidal grainstone to packstone (B), whole-fossil wackestone (C), pelletoidal to evaporitic wackestone to mudstone (D), mixed terrigenous carbonate (E), bioturbated sandstone (F), and cross-bedded sandstone (G). Note that the diagonal of this transition table consists of structural zeros because rock A cannot undergo transition to rock A, rock B cannot undergo transition to rock B, and so forth. Also, the transition table was sparse, with 24 of the 49 cells having fewer than five observed events.

Driese and Dott analyzed these data using log-linear analysis: "a stepwise selection procedure was used for identification of positive and negative departures from facies transition frequencies expected under a quasi-independent model" (p. 588). Their analysis identified eleven cells of the transition table for which the difference between observed and expected frequencies was statistically significant (evaluated using chi-square, least significant $p < 0.088$). Recall that each of these eleven cells indicates a specific transition of the form: rock type i is followed by rock type j. Two patterns of transitions defined in terms of these eleven cells were summarized via flowcharts. One chart considered transitions observed *more often* than expected given chance, and the other chart considered transitions observed *less often* than expected given chance. The more often seen pattern was A leads to C, which in turn leads to D (and to B), which in turn leads to B (and to E), which in turn leads to E (and to D). The less often seen pattern was B leads to A, and both A and C lead to both F and G. Considered together, these two patterns—and eleven assignment rules—correctly identified 178 of the total of 514 events in the transition table: Overall percentage accuracy in classification (PAC) = 34.63%.

These data were analyzed as a nondirectional categorical ODA: Rows corresponded to the rock at state i (the multicategory class variable), and columns to the rock at state $i + 1$ (the polychotomous attribute). Weighting by prior odds was used because there were different numbers of rocks of different types, and LOO validity analysis is used to assess model stability. The analysis was accomplished using the following ODA script.

```
OPEN ex111.dat;
OUTPUT ex111.out;
CATEGORICAL ON;
TABLE 7;
CLASS ROW;
LOO;
MC ITER 10000;
GO;
```

The ODA model can be read directly from the output report. Note that when data are entered using tabular format, ODA software uses numerical codes to identify categorical variables. In the present example, row and column A are coded using 1, row and column B are coded using 2, and so forth (row and column J are coded using 10—the largest categorical variable that can be analyzed by ODA software). The alphanumerically translated ODA model was, If column (i.e., attribute) = A, then row (i.e., class) = F; if column = B, then row = D; if column = C, then row = A; if column = D, then row = C; if column = E, then row = B; if column = F, then row = G; and, finally, if column = G, then row = E. The nature of these individual assignment rules—and of the sequential structure that they imply when they are considered as a set—are both more easily understood when summarized using symbolic notation.

Consider the first assignment rule: If column (attribute) = A, then row (class) = F. To flowchart this rule, imagine that you will use this rule to classify data. If the attribute of the observation to be classified has the value A, then, by this rule, that observation is classified as being a member of Class F. We find it useful to think of this in the terms "A (attribute) leads to F (class)." This rule may be summarized symbolically using the notation

$$A \rightarrow F,$$

where the letters identify rocks and the arrow gives the direction of the classification—that is, from attribute to class. In this manner, all six ODA-derived assignment rules for this example (versus eleven assignment rules for the Markov models) may be summarized symbolically:

$$A \rightarrow F, B \rightarrow D, C \rightarrow A, D \rightarrow C, E \rightarrow B, F \rightarrow G, \text{ and } G \rightarrow E.$$

The construction of a flowchart that summarizes these assignment rules simply involves connecting all six assignment rules, ensuring that the connections are logically consistent. Of course, it is necessary to select one of the rules to begin the flowchart. The implication of the decision concerning which assignment rule to select in order to begin a flowchart has not yet been systematically studied. Clearly, this decision influences the identity of the first-appearing (left-most) and last-appearing (right-most) component in the flowchart, but it does not influence the pattern of the mapped sequential structure. For example, starting the flowchart with the first assignment rule illustrated above would create a sequence beginning with rock A. In contrast, starting the flowchart with the last assignment rule illustrated above would create a sequence beginning with rock G.

In the absence of theory that otherwise directs one's choice, a starting heuristic must be devised in order to guide the selection of an assignment rule with which to begin the flowchart. Possible starting heuristics include *random* (select an initial assignment rule randomly); *actual* (select the initial assignment rule that reflects the transition with which the data—that is, the empirical sequence—actually started); *fastest* (select the initial assignment rule that reflects the transition that occurs with greatest frequency in the transition table—that is, the fastest-occurring transition); or *slowest* (select the initial assignment rule that reflects the transition that occurs with least frequency in the transition table—that is, the slowest-occurring transition).

For example, using the fastest starting heuristic in the present case, the rule "C leads to A" is selected to initialize the flowchart because it is associated with 52 correctly classified transition events—more than any of the other five assignment rules. To construct the flowchart, begin by writing the first assignment rule:

C → A .

Because this rule ends with A, and, by examining the five remaining assignment rules, it is seen that A leads (is classified) into F, therefore append the flowchart accordingly:

C → A → F .

Similarly: F leads to G,

C → A → F → G ;

G leads to E,

C → A → F → G → E ;

E leads to B,

C → A → F → G → E → B ;

and B leads to D,

C → A → F → G → E → B → D .

As is required of a logically consistent flowchart, the assignment rule "D leads to C" may be indicated using an arrow pointing backward from D into C. Using the present notation however, unless otherwise noted it is assumed that the right-most component of the sequence (here, D) leads back into the left-most component of the sequence (here, C), which then recycles or repeats itself. In this example the flowchart, translated below, clearly indicates that the ODA model represents a one-dimensional sequentially ordered structure (process) in the state transition table.

Whole-fossil wackestone	→	Bioclastic grainstones to packstone	→	Bioturbated sandstone	→	Cross-bedded sandstone	→	Mixed terrigeneous carbonate	→
		→ Pelletoidal grainstone to packstone		→ Pelletoidal to evaporitic wackestone to mudstone		→ Whole-fossil wackestone		→ ...	

This ODA model correctly classified 217 of the total of 514 events in the transition table: Overall PAC = 42.2%, versus 34.6% for the log-linear models. In this example, the ODA model was more parsimonious—requiring 45% fewer model terms than the log-linear models (6 versus 11 assignment rules, respectively), and yet it also was more powerful—yielding a 22% increase in overall PAC. Moreover, the ODA model is sensible geologically, reflecting the effect of gradually fluctuating depths (or energy levels) on deposition in a subaqueous environment (Jack Yarnold, personal communication, 1996).

Classification Performance Index	Value (%)
Overall classification accuracy	42.22
Effect strength, sensitivity	32.00
Effect strength, predictive value	38.92
Total effect strength	35.46

Note. Here, sensitivity is the accurate classification of rocks *in* state i, and predictive value is the accurate classification of rocks *into* state $i + 1$.

Finally, the classification performance of the ODA model, stable in LOO validity analysis, was moderate in practical terms (total effect strength = 35.46%) and statistically significant under the null hypothesis that rocks at state i cannot be discriminated on the basis of rocks at state $i + 1$ (estimated $p < 0.0001$).

Analyzing Turnover Tables

Assignment of observations into one of two or more mutually exclusive and exhaustive types is pervasive in the social and medical sciences. In both longitudinal and cross-sectional research involving such qualitative measures it is a common practice to assess observations' types twice—with testings separated by either a short or a long time period—in order to determine the extent to which typological assignments are temporally stable as well as whether any notable transitions in typological assignments might occur across testings. Because such data are usually inappropriate for analysis by traditional statistical methods, researchers often simply report eyeball-based summaries of the cross-classified data.

In general, *turnover tables* are used to summarize the behavior of a categorical variable assessed for a sample of observations measured at two points in time. They are fundamental in the panel analysis of discrete data and have been extensively studied (Bishop, Fienberg, & Holland, 1975; Haberman, 1979; Hagenaars, 1990). A turnover table actually constitutes a special case of a state transition table, in which rows indicate observations' responses or types on the categorical variable at the first recording, time 1 (i.e., the state at time $i-1$), and columns indicate observations' responses or types at the second recording, time 2 (i.e., the state at time i). If observations' responses to the categorical variable (or type) are completely consistent at time 1 and time 2, then all of the data will fall in the diagonal of the table, and the test–retest reliability is perfect (in this circumstance the turnover table will necessarily be sparse). In contrast, as observations begin to change their response across testings (or their type changes), the temporal reliability of the variable decreases. Analysis of turnover tables thus requires assessing both the *stability* (on the diagonal) and *instability* (off the diagonal) of a categorical variable. Conventionally analyzed using log-linear analysis or linear structural relationships, analysis may be problematic when the turnover table is sparse or involves structural or empirical zeros (cf. Bakeman & Quera, 1995; Bishop et al., 1975; Gilbert, 1993; Haberman, 1979; Hagenaars, 1990).

Example 11.2

As an example of the use of ODA to assess stability for a turnover table in which consecutive codes are allowed to repeat, consider Loo's (1996) data on the temporal stability of learning styles. Kolb's (1984) experiential learning model posits that learning involves a four-stage cycle beginning with concrete experience, followed in turn by observation/reflection, assimilation of concepts/construction of generalizations, and new interactions with the world. Learning styles are hypothesized to be relatively stable, but change (e.g., development) is also expected. A measure of learning style, Kolb's (1984) revised Learning Style Inventory (LSI-1985) provides scores on two independent dimensions—concrete experience/abstract conceptualization and active experimentation/reflective observation, which are crossed to form quadrants representing four theoretical learning styles: accommodator, diverger, assimilator, and converger. Loo (1996) administered the LSI-1985 at the beginning and again at the end of the semester (10-week inter-test separation) to 149 undergraduate students.

Time 1	Time 2			
	Assimilator	Accommodator	Converger	Diverger
Assimilator	26	12	10	9
Accommodator	4	12	4	8
Converger	7	3	16	6
Diverger	4	6	7	15

Because the table is sparse (25% of the cells have four or fewer observations) these data were not amenable to traditional analysis, and Loo provided an eyeball data analysis:

As seen in the diagonal entries . . . the largest effect is the stability (42.8% to 50.0%) of learning styles for each of the four styles. The direction of change indicates that Assimilators who change tend toward Accommodators (21.1%), Accommodators tend toward Divergers (28.6%), Convergers tend toward Assimilators (21.9%), and Divergers tend toward Convergers (21.8%). (pp. 531–532)

ODA analysis of stability in turnover tables involves attempting to classify observation type at the second testing directly on the basis of observation type at the first testing. In the present context, it is hypothesized that learning styles are stable. That is, an observation's learning style at time 2 (multicategory class variable) is hypothesized to be identical to the observation's learning style at time 1 (polychotomous attribute). Thus, the data are hypothesized to fall in the major diagonal of the turnover table. Weighting by prior odds is appropriate because of class category sample size imbalance. LOO analysis would be superfluous. The analysis was accomplished using the following ODA script.

```
OPEN ex112.dat;
OUTPUT ex112.out;
CATEGORICAL ON;
DEGEN ON;
TABLE 4;
CLASS COL;
DIRECTIONAL < 1 2 3 4;
MC ITER 10000;
GO;
```

The directional ODA model for the stability hypothesis correctly classified 69 of the total of 149 observations, yielding an overall classification accuracy of 46.3%, and estimated $p < 0.0001$. The total effect strength of 27.97% indicates that this is a moderate effect in practical terms.

Classification Performance Index	Value (%)
Overall classification accuracy	46.31
Effect strength, sensitivity	27.50
Effect strength, predictive value	28.45
Total effect strength	27.97

These results suggest that learning styles as assessed via the LSI-1985 instrument show a statistically significant, practically moderate degree of temporal stability. Nevertheless, the majority of the entries in the turnover table (53.7%) are misclassified by the stability model.

Do these residual data simply reflect random error, or does a statistically reliable transition profile underlie the residual data? We shall return to this question in the following chapter.

Example 11.3

As an example of the use of ODA to assess stability for a turnover table in which consecutive codes are *not* allowed to repeat (for which the major diagonal of the turnover table consists of structural zeros), consider Foa's (1971) data on the exchange of material and psychological resources. Theoretically, six resource categories (love, services, goods, money, information, status) are cyclically ordered in the two-dimensional space created by crossing *particularism* and *concreteness*, and categories near each other in the cyclic ordering (e.g., love and status) are more similar than further separated categories (e.g., love and information). In the schematic illustration of Foa's dissimilarity hypothesis, double arrows indicate the hypothesized most *dissimilar* pairings of resource categories (see Figure 11.1).

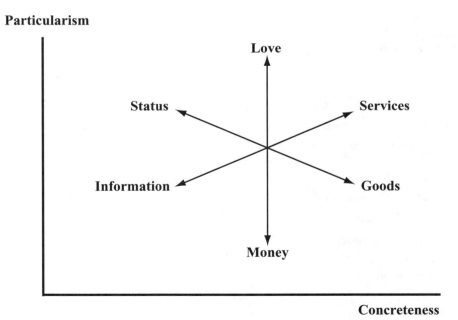

Figure 11.1. Foa's (1971) theoretical model on the exchange of material and psychological resources.

To test the dissimilarity hypothesis, 37 people received three messages for each resource category: Separately for each received message, a person returned a message—selected from an accompanying deck—judged to be most unlike the message received. In the deck all categories were represented except the category from which the message was received. The diagonal was thus structural zeros. Bishop et al. (1975) note that receipt of the message may be conceived as having occurred at time 1, and return of the message may be conceived as having occurred at time 2 (stimulus and response, respectively). The resulting turnover table follows.

Message Received (Time 1)	__Love__	__Status__	__Information__	__Money__	__Goods__	__Services__
Love	0	5	21	48	29	8
Status	4	0	19	27	30	31
Information	20	11	0	20	25	35
Money	56	10	21	0	4	20
Goods	42	18	27	6	0	18
Services	12	20	37	26	16	0

(columns under heading: Message Returned (Time 2))

Note that although row marginals were constrained to be equivalent, column marginals were (and are) not. And, in addition to the structural zeros in the diagonal, matters were further complicated by two cells having fewer than five entries. Accordingly, Foa presented an eyeball analysis of the turnover table: "the highest frequency occurred in the cell three steps removed from the diagonal with a decrease as one approached the diagonal from either direction" (p. 347).

Testing the dissimilarity hypothesis via ODA involves attempting to classify the message returned (second testing) on the basis of message received (first testing). In the present context, it is hypothesized that the pairings are most dissimilar. Thus, if a love (row = 1) message is received, then money (column = 4) is the hypothesized least similar response (and 4 is the first code in the DIRECTIONAL command). If a status (row = 2) message is received, then a goods response (column = 5) is hypothesized (and 5 is the second code in DIR). Finally, if a services (row = 6) message is received, then an information response (column = 3) is hypothesized (and 3 is the last, or sixth, code in DIR). Prior odds weighting is appropriate because of class (column) category sample size imbalance. LOO analysis would be superfluous. The p estimated by ODA is slightly conservatively biased (true $p \leq$ estimated p): Use the adjust keyword of the MCARLO command to remove this small bias. The analysis was accomplished using the following ODA script.

```
OPEN ex113.dat;
OUTPUT ex113.out;
CATEGORICAL ON;
TABLE 6;
CLASS COL;
DIRECTIONAL < 4 5 6 1 2 3;
MC ITER 10000;
GO;
```

The results suggest statistically significant but practically weak support for the dissimilarity hypothesis. The a priori ODA model for the dissimilarity hypothesis correctly classified 224 (33.6%) of the total of 666 entries: This result was statistically significant ($p < 0.0001$) but weak in practical terms (effect strength for sensitivity = 19.48%; effect strength for predictive value = 20.36%; total effect strength = 19.92%).

Autocorrelation (Time-Series) Analysis

Time-series models are widely used to study processes underlying sequentially ordered data, in fields such as advertising, criminal justice, economics, education, earth science, finance, medicine, political science, psychology and sociology (e.g., Barlow & Hersen, 1984; Box & Jenkins, 1976; Glass, Willson, & Gottman, 1975; Hartmann et al., 1980; Jenkins & Watts, 1968; McCleary, Hay, Meidinger, & McDowall, 1980; McDowall, McCleary, Meidinger, & Hay, 1980; Pindyck & Rubinfeld, 1976). Consistent with the Markov model approach, when conducting time-series analysis it is important to distinguish between *stationary processes*, which are invariant over the course of the sequence, and *nonstationary processes*, which change over the course of the sequence. When time-series data are stationary it is a relatively straightforward procedure to develop a model, involving fixed coefficients, that seeks to explain sequential processes in the data. When time-series data are nonstationary, the models are more complex and require that increasingly restrictive assumptions are satisfied by the data in order for the results of the analysis to be valid. There is debate concerning the degree to which there is evidence of nonstationarity in real-world data, the degree to which such nonstationarity presents a problem for time-series models, and what to do in situations where there is a nonstationarity problem (e.g., Gorsuch, 1983; Huitema, 1985; Wampold, 1988).

How does one go about determining whether a time-series—that is, a regularly-spaced stream of consecutive sequentially ordered data points—reflects a stationary process? By definition, a stationary process is fully determined by its mean, variance, and *autocorrelation function* (ACF). The ACF provides an index of the degree of *serial dependency* that exists between neighboring data points in a time-series. To understand how the ACF functions, one must be familiar with the notion of a *lag*, which we pause to review.

To understand the notion of a lag, imagine that we wished to investigate a sequential process over some specific period of time measured forwards (prospectively) or backwards (retrospectively) with respect to the present. The "clock" with which the passage of time is measured is set in a manner so that it "ticks" (i.e., indicates the passage of one time unit) once every time that a measurement—corresponding to an individual element in the time-series—is recorded. Time begins at zero, and the clock begins to run when the study begins. When the clock indicates that the first time unit has elapsed (time = 1) the first measurement is recorded; when the clock indicates the second time unit has elapsed (time = 2) the second measurement is recorded, and so forth. Thus, the time period selected to serve as the study time domain corresponds directly to some discrete number of time units, and therefore corresponds directly to the number of discrete measurements.

Imagine conducting a time-series experiment: Set the clock, begin the study, and collect data over the period of the study. At the completion of this procedure, one obtains a time-series consisting of T sequential data points, with the sequential order of each data point indicated using an index number ranging between 1 (the first data point in the series, recorded at time = 1) and T (the last data point in the series, recorded at time = T): $x1, x2, \ldots xi, \ldots xT$. Because the index number for each data point is the same as the time that data point was recorded, there is no "lag" between the time-series and the clock. Thus, this is referred to as a lag-0 time-series. Suppose, however, that we began to collect data when the clock registered 1, rather than 0 as was the case for a lag-0 series?

In a lag-1 time-series, the data collection begins after the clock has already indicated the lapse of one time unit. Thus, rather than recording the first data point, $x1$, when the clock registers time = 1, for a lag-1 time-series we record the first data point, $x1$, when the clock registers time = 2. And, if we still intended to collect T measurements, the last measurement would be recorded at time T + 1 for a lag-1 time-series. Of course, one cannot record a measurement at time T + 1 because, by definition, only T measurements are made. Thus, for a lag-1 time-series, only T − 1 measurements are possible. In general, for a lag-k time-series, the first observation is recorded at time k + 1, and only T − k measurements are possible.

For any time-series, ACF(k) represents the correlation (i.e., autocorrelation) between pairs of corresponding measurements, matched on the basis of their index numbers, from a lag-0 time series and its kth lag. Thus, ACF(1) represents the autocorrelation between T − 1 pairs of measurements, matched on the basis of their index numbers, from a lag-0 and its corresponding lag-1 time-series. Similarly, ACF(2) is calculated by correlating the T − 2 pairs of matched measurements from a lag-0 and its lag-2 time-series, and ACF(k) is calculated by correlating the T − k pairs of matched measurements from a lag-0 and its lag-k time-series.

In applied research it is customary to compute the first T / 4 ACF(k), and then to plot ACF(k) against k, for k = 1, 2, ... T / 4 (this plot is called a *correlogram*). ACF(k) is used to assess both the degree of serial dependency in, and the stationarity of, a time-series. High values for ACF(k) estimates (because these are correlations, their theoretical maximum absolute value is one) are interpreted as evidence of serial dependency in the time-series. When the slope of the correlogram drops off quickly as k increases, this is considered evidence that the time-series is stationary. However, when the slope of the correlogram does not drop off sharply as k increases, this is considered evidence of nonstationarity in the time-series.

ODA may be used to analyze categorical data that are problematic for traditional time-series methods. Traditionally, ACF(k) constitutes a (nonparametric) correlation coefficient that may be used to estimate the proportion of shared variance between two arrays of T − k measurements separated by k time units. This proportion of shared variance is interpreted as an index of the serial dependency in the data. ODA, in contrast, determines the PAC achieved using the value of a measurement at time T − k to predict the value of the measurement at time T, and thus speaks directly to the issue of how accurately one can predict the value of xT on the basis of the value of xT − k. The overall PAC (and effect-strength measures) achieved using the ODA model is then interpreted as an index of the serial dependency in the data.

To illustrate the ODA analogue to ACF(k) for categorical data, consider the simplest case of a binary attribute. The first step in the ODA approach involves transforming the time-series data into a state transition table (see Whitehurst, Fischel, DeBaryshe, Caulfield, & Falco, 1986, for discussion concerning the use of kappa versus lag-1 autocorrelation in the analysis of state transition matrices). In this transition table, rows correspond to the state (or value) of the measurement at time *i*, and columns correspond to the state of the measurement at time *i* − k, where k is the desired lag. We should note that the conventional approach for computing ACF(k) for binary data involves use of the lag-k phi coefficient, where phi is the correlation between binary measures (Gardner, Hartmann, & Mitchell, 1982). Reynolds (1977) discusses the limitations of phi when the relative frequency of 1s and 0s in the time series is imbalanced (see also Davenport & El-Sanhurry, 1991).

Example 11.4

As an example of the use of ODA to assess serial dependency in time-series data reflecting a binary attribute—the ODA analogue to ACF(k)—imagine that a batting coach desired to evaluate the degree of serial dependency in a lag-1 time-series consisting of all sequential pitches made by the pitcher of the opposing team during the first two innings of a World Series baseball game. Each pitch was coded as being either a strike (S) or a ball (B). (This is not an ideal time-series because, in defiance of the assumption to the contrary, pitches may not come at regularly spaced intervals. The degree to which this assumption is violated can be limited to some extent by the sensitivity or metric of the clock adopted to assess the passage of time. However, situations with runners on base may be so well discriminable from situations with the bases empty, in terms of time taken per pitch, that *interrupted time-series* methods may be appropriate [McDowall, McCleary, Meidinger, & Hay, 1980].) Imagine that twenty pitches were recorded:

S, S, B, S, B, S, B, S, S, S, B, B, S, B, B, S, S, B, B, S ;

where the time-series begins at the left and proceeds in time sequentially toward the right. To transform this series into a state transition table, it is necessary to work from right to left as one compares each sequential pair of measurements, and to step backward one time period at a time. For example, consider the rightmost pair of measurements. The rightmost measurement (S) is treated as state i, and the second measurement from the right (B) is treated as state $i-1$ (call these the 20th and 19th measurements, respectively). The *autoregressive transition* implied by comparing the 20th (S) and 19th (B) measurements is, "S (at state i) leads to B (at state $i-1$)." Comparing the 19th (B) and 18th (B) measurements, the implied autoregressive transition is, "B leads to B" (the diagonal of the autoregressive state transition table usually is not be composed of structural zeros). As shown below, all T − 1 = 19 pairs of [xi, x($i-1$)] measurements may be evaluated and tallied in the appropriate cell of the autoregressive state transition table.

State at Time I: xi	State at Time $i-1$: x($i-1$)	
	S	B
S	4	6
B	6	3

The transition table is then analyzed as a categorical ODA with row treated as the class variable because we wish to determine the extent to which measurements at time $i-1$ may be used to accurately classify the measurements at time i. Weighting by prior odds is appropriate because of the different number of strikes and balls. The analysis is nondirectional (the null hypothesis is pitches at state i cannot be discriminated on the basis of pitches at state $i-1$): The nature of any serial dependency is an empirical question. The following ODA script was used for this analysis.

OPEN ex114.dat;
OUTPUT ex114.out;
TABLE 2;

CATEGORICAL ON;
CLASS COL;
PRIORS;
MCARLO ITER 10000;
GO;

The alphanumerically translated ODA model was, If column (attribute) = S, then predict that row (class) = B; and if column = B, then predict that row = S. Symbolically

$S \rightarrow B$, and $B \rightarrow S$.

Next, these two ODA-derived assignment rules were organized using a flowchart with the exact starting heuristic (see Example 11.1). Examination of the original time-series revealed that the series began with (i.e., the initial measurement was) an S, and therefore so did construction of the flowchart.

$S \rightarrow B$.

Using the notation of Example 11.1, this flowchart is consistent and complete. This ODA model represents a one-dimensional sequentially ordered structure (process) in the autoregressive state transition table: The pitcher is alternating strikes and balls. This model correctly classified 12 of the total of 19 events in the transition table (overall PAC = 63.16%). This moderate level of classification accuracy was weak (total effect strength = 26.67%) and not statistically significant (exact $p < 0.37$). In summary, the ODA model was unable to achieve statistically or practically significant levels of classification accuracy using a lag-1 attribute, indicating that the opposing pitcher's pitches were not lag-1 serially dependent—that is, did not have a significant underlying ACF(1) effect—in the first two innings.

In addition to examining ACF(1), the correlogram—a plot of the first T / 4 ACF(k) versus k—is also examined in order to determine whether the time-series is stationary. If the series is stationary, then the slope of the correlogram should drop rapidly over increasing values of k. ODA may be used in this context, by plotting overall PAC (or standardized effect strength) against k for each of the first T / 4 ACF(k).

Using ODA, the lag-2 autocorrelation, ACF(2), analogue is computed in exactly the same manner as ACF(1), except that the columns in the autoregressive state transition table correspond to measurements taken at state $i - 2$, rather than at state $i - 1$ as is true for the lag-1 problem. And the ACF(k) estimate of serial dependency is computed by using columns corresponding to state $i - k$. In the present example, because there are 20 measurements, it is appropriate to compute the first 20 / 4 = 5 ACF(k) models. For example, the autoregressive state transition table for the ACF(5) analysis, which has T − k = 20 − 5 = 15 events, has the following form:

	State at Time $i-5$: $x(i-5)$	
State at Time I: xi	S	B
S	3	5
B	5	2

In this manner, the ACF(1), ACF(2), ACF(3), ACF(4), and ACF(5) estimates were obtained using ODA (this is not illustrated), and the following results were obtained.

Overall PAC for ACF(1) = 63.16%
Overall PAC for ACF(2) = 66.67%
Overall PAC for ACF(3) = 52.94%
Overall PAC for ACF(4) = 68.75%
Overall PAC for ACF(5) = 66.67%

Here, a correlogram (actually, a *PACogram*) need not be plotted—it may be imagined because eyeball inspection clearly reveals that overall PAC drops precipitously only for ACF(3): All other estimates were *greater* than ACF(1). Thus, over the first two innings, the series of pitch selections by the opposing pitcher reflects a stationary process for runs of at least three consecutive pitches.

What if the measurement taken at each recording had been polychotomous, involving more than two discrete content codes? For example, this could involve coding type of pitch (curve, fastball, slider), or coding both type and location of pitch (inside or outside curve; inside or outside fastball; inside or outside slider). In that case, the analysis would unfold and be interpreted in a manner analogous to the example just described, but techniques appropriate for applications involving a multicategory class variable and a polychotomous attribute would be used to identify ODA models. And, of course, ODA may similarly be used to obtain ACF(k) for applications involving ordinal data with ten or fewer discrete levels.

Repeated Measures (Within-Subjects) Analysis

In analysis of variance (ANOVA), main and interaction effects may be purely *between-subjects*, involving the comparison of two or more independent groups of observations with respect to *a single measurement* of an attribute (the citations provided for ANOVA when randomized block designs were discussed are applicable here as well). We have already described the use of ODA to analyze data for between-subjects applications involving two (binary class variable) or more (multicategory class variable) groups and either an ordinal or continuous (interval or ratio) attribute. It should be noted that considering a single attribute at a time (i.e., per analysis) constitutes an ANOVA or UniODA problem, whereas considering multiple attributes simultaneously constitutes a multivariate analysis of variance (MANOVA) or MultiODA problem. Although we will discuss nonlinear hierarchically optimal MultiODA models (chap. 12, this volume), linear MultiODA models involve substantial intensive mathematics and lie beyond the accompanying ODA software (see epilogue).

In ANOVA, main and interaction effects may also be purely *within-subjects*, involving the comparison of *multiple (repeated) measurements* of an attribute for a single group of observations (e.g., see ANOVA citations). Such designs are the focus of the present discussion. Within-subjects analysis must contend with inherent dependence in the data—because the same observation is providing repeated assessments, the assessments are correlated—that does not exist in between-subjects analysis.

When analyzing repeated measures designs, it is important to remember that the p value that ODA provides is *biased*. That is because the theory used to generate p does not account for the dependence that is inherent in within-subjects data. That is, ODA distribution theory assumes *independent groups*, whereas in within-subjects designs the groups are *dependent*. Fortunately, the nature of this bias is *conservative*. For example, imagine a single group tested twice on a single attribute: ODA will attempt to find a cutpoint that discriminates scores at one testing from scores at the other testing. However, because the data are from the same people, the scores at the two testings will thus be correlated and more difficult to discriminate than would be true if the data were instead from two independent groups of people. Thus, when data are correlated (dependent) it is more difficult to obtain a given level of classification performance than when the data are orthogonal (independent). Because it would be more difficult to achieve the observed level of classification performance using dependent groups, the p associated with the observed performance would be lower ("more significant") if distribution theory had assumed dependent groups. It is better that estimated p is conservatively versus liberally biased (over- versus under-estimated, respectively) in situations where an effect is judged to be statistically significant. It is more ambiguous when an effect is judged to be nonsignificant, particularly when the significance criterion is narrowly missed. Using effect strength measures to assess the practical significance of a result is a particularly attractive recourse when p is known to be biased.

Finally, in ANOVA, interactions may also reflect *mixed effects*, involving the product of a between- and a within-subjects factor. Mixed effects may be included in complex suboptimal multiattribute classification models that are subsequently optimized (chap. 9, this volume), or as is discussed ahead, may be analyzed via GenODA-based or classification tree-based models.

Example 11.5

As an example of the use of ODA to analyze data reflecting a single-factor within-subjects design—which conventionally would be analyzed using within-subjects t tests and one-way repeated measures ANOVA, consider Wyte et al.'s (1996) prospective study of the efficacy of lecture- and case-study-based teaching methods in an emergency medicine residency program. On three different occasions—the first day of the residency (baseline), midway through the residency (midterm), and the final day of the residency (final)—a total of 63 residents completed parallel forms (of equal length and comparable difficulty) of a standardized test assessing knowledge in nine fundamental facets of emergency medicine. Of primary interest in this study is the validity of the directional alternative hypothesis that scores will increase over the course of the residency: Scores are hypothesized to be lowest at baseline, highest at final, and intermediate at midterm (the null hypothesis is that this is not true). Of secondary interest is the GenODA nondirectional hypothesis that this finding will generalize across gender (1 = male, 2 = female). Presently this analysis requires that data for a single observation are coded as though they actually were from observations from three different groups, dummy-coded using 1 (for baseline scores), 2 (for midterm scores), and 3 (for final scores). Thus, data for each observation have been entered in free format in the following order, that is designed to imitate the data of three observations: gender code, testing (i.e., "group") code (1), baseline score (and thus the "first observation" is constructed), gender code, testing code (2), midterm score (the "second observation"), gender code, testing code (3), and final score (the "third observation").

From the perspective of ODA there are not 63 observations, but rather there are 3 × 63 = 189 observations. All missing data are indicated using "–9". Four residents were missing midterm scores, and two were missing final scores. Thus, for ODA there were 189 – 6 = 183 observations (because class category sample sizes were different, prior odds weighting was used). In contrast, for conventional statistics observations with any missing data are deleted from analysis (or missing data are estimated). Presently, with conventional case-wise deletion there would only be apparent data for (63 – 6) × 3 = 171 observations. The omnibus analysis comparing score across all three testings was conducted using the following ODA script.

```
OPEN ex115.dat;
OUTPUT ex115.out;
VARS sex testing score;
CLASS testing;
ATTRIBUTE score;
DIRECTIONAL < 1 2 3;
LOO;
MCARLO ITER 10000;
MISSING ALL (-9);
GO;
```

The ODA model was, If score ≤ 12.5, then predict baseline testing; otherwise, if 12.5 < score ≤ 15.5, then predict midterm testing; otherwise, if score > 15.5, then predict third testing. Training classification performance achieved by this model was moderate in practical terms (total effect strength = 30.48%), statistically significant (estimated $p < 0.0001$), and stable in LOO validity analysis. These findings offer statistically significant, practically moderate support for the a priori hypothesis that scores increased over training. If one were interested in conducting planned comparisons, or all possible comparisons, to further isolate the foci of this effect, then analysis would proceed as illustrated in Example 6.5.

The secondary objective of this study was to assess whether the effect just notes generalized across sex. Accordingly, in order to determine whether data should be normatively standardized separately by sex (i.e., because of the possibility of paradoxical confounding), the following ODA script was used to assess whether gender could be discriminated on the basis of score.

```
CLASS sex;
DIRECTIONAL OFF;
GO;
```

This resulting ODA model yielded classification accuracy that was stable in LOO analysis, but that was weak in practical terms (total effect strength = 11.2%) and was not statistically significant (estimated $p < 0.55$). Accordingly, it appears that the use of raw data is appropriate. The GenODA model used to assess the generalizability of the directional alternative hypothesis that scores increased over the course of residency training was accomplished by appending the following ODA script.

```
GEN sex;
GO;
```

Because the GenODA model was identical to the ODA model, the classification performance and p achieved using the GenODA model for the pooled sample is identical to the classification performance and p achieved using the (identical) ODA model for the (identical) total sample, for both training and LOO analyses.

However, the GenODA analysis further reveals that this model generalizes for males (total effect strength = 33.70%, estimated $p < 0.0001$) versus females (total effect strength = 27.05%, estimated $p < 0.0001$). These findings offer statistically significant, practically moderate support for the a priori hypothesis that scores increased over training in a comparable manner for males and females (note this is a mixed-effect hypothesis). As before, if one were interested in conducting planned or all possible comparisons to further refine the exact nature of this effect, then analysis would proceed as illustrated in Example 6.5.

This example focused on comparing the responses of a single group of observations to *a single attribute assessed across two or more testings* (see also Levenson, Grammer, Yarnold, & Patterson, 1997). An alternative focus involves comparing the responses of a single group of observations to *two or more different attributes assessed at a single testing*. Here, each attribute is treated as a "testing." The nondirectional alternative hypothesis (directional hypotheses are also possible) is that groups (i.e., attributes) can be discriminated on the basis of score: The null hypothesis is that this is not true (Russell, Bryant, & Estrada, 1997, use a conceptually similar approach to compare bootstrap distributions of goodness-of-fit statistics for non-nested P-technique structural models). Of course, in order for this analysis to be meaningful it is imperative that the attributes are measured on a common scale. If attributes are not measured using a common scale, then an interactive transformation of the attributes is required prior to conducting the analysis.

Single-Case (N-of-1) Analysis

Finally, we briefly discuss several strategies that may be used to conduct statistical evaluation of single-case (case-study) data. Imagine that a single observation has been scored on a single attribute on multiple occasions: For example, we may measure the temperature of a hospitalized patient every hour; the closing price of a stock every day; the number of petitions for unemployment insurance each week; or the number of handgun fatalities every year. Given such a series, we may wish to statistically assess whether the data values are higher (or lower) before a particular intervention (or event) occurred. For example, the patient might undergo an intervention against infection, and the physician might wish to know if temperature recordings for the first 6 hours after the intervention are lower than recordings for the 12 hours immediately before the intervention. Or, we may wish to assess if the number of handgun fatalities is lower for the years following the passage of the Brady Bill, versus the most recent fifteen years preceding the Bill.

When only a few data points—such as ten or fewer—are available, statistical methods based on classical test theory may be best suited (e.g., Choy et al., 1995; Mueser, Yarnold, & Foy, 1991; Nishith, Hearst, Mueser, & Foa, 1995; Yarnold, 1988, 1992). For example, to study treatment effectiveness for a sample of patients, Yarnold, Feinglass, Martin, and McCarthy (1997) initially evaluate data on individual patients. For example, their approach first determines whether each individual patient has a greater post-test score (improved functional status), a greater baseline score (diminished functional status), or comparable baseline and post-test scores (stable functional status). Once the outcome status of individual patients is known, then ODA-based comparison of distributions of outcomes between, for example, samples of control versus intervention patients may be made in terms of the relative proportion of patients showing improvement. Thus, rather than conceptualizing outcomes in terms of change in mean scores across time for samples of patients, "pre-processing" the data via single-case analysis allows one to conceptualize outcomes in terms of the relative probability that a patient will benefit from the intervention. In the latter approach there is no consideration of mean scores attained by samples.

However, for applications involving a larger number of measurements, ODA is a powerful single-case statistical analysis (however, recall the caveat about the assumption of independence and the conservative bias in p that occurs when data are serially dependent). For example, imagine that a person completed a questionnaire that had 20 items measuring one factor and 11 items measuring the opposite pole of that factor (e.g., Type A versus Type B behavior, respectively). By dummy-coding the Type A items as "Group 1 observations," and the Type B items as "Group 2 observations," ODA may be used as for a repeated measures design to determine whether group (i.e., pole of the factor) can be discriminated by score on the attribute (i.e., item rating). If, for example, a person's responses to Type A items is greater than responses to the Type B items, then it may be inferred that the person is more like a Type A than like a Type B.

Alternatively, rather than being separate ends of a single factor, these may be different factors—for example, instrumental versus expressive interests. In a parallel manner, ODA may be used to assess whether group (i.e., type of factor) can be discriminated by item rating. For example, a person's scores on items assessing instrumental interests may be greater than scores on items assessing expressive interests, indicating that the person is significantly more interested in instrumental activities. Of course, both of these procedures assume that attributes have been measured using the identical metric. In cases in which this is not true, interactive transformations may be used to equate scale metrics.

Finally, imagine an application in which the long-term effectiveness of a drug therapy is being evaluated for people with severe allergies, who require a variable amount of drug therapy (e.g., via inhaler) depending on their symptoms. Much of the time no or very little drug is required, and other times, acutely high dosages are required. Imagine that we had a daily record of the dose taken for a single patient, for a period of 4,983 days. Then, we have a similar record for the same patient, for the 179 days changing from the original drug to a new synthetic equivalent. The issue is whether the amount of the new drug required is less than the amount of the old drug. In this case, paralleling the prior examples, code the original drug as "Group 1," the new drug as "Group 2," and use ODA to determine whether group (drug) can be discriminated on the basis of dose.

CHAPTER 12

Iterative Decomposition Analysis

Conventional statistical procedures designed for applications involving multiple attributes often use iterative analyses, in which successively or hierarchically added terms add incrementally to the overall goodness-of-fit of the "omnibus" model. For example, multivariable methods such as stepwise multiple regression allow independent variables to enter the model until they fail to meet a user-specified criterion, typically expressed in terms of p, incremental change in R^2, and/or the number of attributes already present in the model. Also, multivariate procedures such as principal components analysis allow factors—linear functions of the original attributes that explain monotonically decreasing portions of the total variance—to enter the model until they fail to meet a user-specified criterion, typically expressed in terms of p, incremental change in R^2, and/or the number of factors already present in the model (e.g., Bryant & Yarnold, 1995). The factors are then interpreted in an effort to identify the structure underlying the correlations. This is reminiscent of the manner in which spectral analysis decomposes a complex wave into a combination of successively determined simple waves (Jenkins & Watts, 1968).

Analogously, iterative use of ODA can identify successive models that explain a monotonically decreasing number of events in a contingency table. Consider, for example, a test–retest reliability table, with rating (1–5) by rater A treated as the rows of the table, and rating (1–5) by rater B treated as the columns of the table. The primary alternative hypothesis is usually that the raters agree: that is, that the data will fall in the major diagonal of the table. But what about any data that do not fall in the diagonal—that is, the off-diagonal data? Are these residual data points randomly dispersed in the off-diagonal cells, or does a reliable pattern underlie them? The methodology that we refer to as iterative (or sequential) ODA-based decomposition of a contingency table (e.g., a state transition table, turnover table, reliability table, etc.) proceeds in the following manner.

Step 1. Initiate the analysis by performing (non)directional ODA on the original table. Refer to this as the *first (or primary) model*—that is, the model that emerged in the first step of the analysis, that correctly classified the greatest overall number of (non)weighted events in the table. If the resulting level of classification performance is judged to be sufficient, then the analysis is completed. If, on the other hand, the resulting classification performance is judged to be unsatisfactory—that is, unacceptably low, then the analysis may appropriately be continued. Proceed to Step 2.

Step 2. Prune the original table—that is, identify the cells that were correctly classified using the ODA model developed in Step 1—and physically change the value in those cells to zero. Because events in these cells have already been correctly classified in Step 1, they do not have to be considered again. Then perform (non)directional ODA on this new pruned table. Refer to this as the *second (or secondary) model*—that is, the model that emerged in the second step of the analysis, and that, except for the model identified in Step 1, correctly classified the greatest overall number of events in the table. If the resulting overall classification performance is judged to be satisfactory, then the analysis is completed. However, if the resulting overall classification performance is judged to be unsatisfactory, then continue to repeat Step 2, each time on the most recently pruned table, until the criterion for success has been achieved.

Several test data sets will help to illustrate how this procedure unfolds. For example, consider the following hypothetical table, which features a one-dimensional linear solution and uniform random residuals. The linear model explains the data in the major diagonal of the initial table. Once the linear ODA has been performed, the table is pruned to change the value of cells already explained to zero. As seen, once the linear model is removed, the remaining non-zero elements in the table are uniformly distributed, and no secondary ODA model is possible.

Initial Table					Table After the First Pruning				
	Class Variable					Class Variable			
Attribute	1	2	3	4	Attribute	1	2	3	4
1	25	5	5	5	1	0	5	5	5
2	5	25	5	5	2	5	0	5	5
3	5	5	25	5	3	5	5	0	5
4	5	5	5	25	4	5	5	5	0

Alternatively, consider the following hypothetical table that features a one-dimensional nonlinear solution and uniform random residuals. The nonlinear model explains the data in (row 1, column 2), (row 2, column 4), (row 3, column 1), and (row 4, column 3) of the initial table. Once the nonlinear ODA has been performed, the table is pruned to change the value of cells already explained to zero. As seen, once the nonlinear model is removed, the remaining non-zero elements in the table are uniformly distributed, and no secondary ODA model is possible.

Initial Table					Table After the First Pruning				
	Class Variable					Class Variable			
Attribute	1	2	3	4	Attribute	1	2	3	4
1	5	25	5	5	1	5	0	5	5
2	5	5	5	25	2	5	5	5	0
3	25	5	5	5	3	0	5	5	5
4	5	5	25	5	4	5	5	0	5

Finally, consider the following hypothetical table, which features a two-dimensional solution involving a primary linear model, a secondary nonlinear model, and uniform random residuals. As in the first example, the first model explains the cells constituting the table major diagonal. The nonlinear model explains the data in (row 1, column 2), (row 2, column 1), (row 3, column 4), and (row 4, column 3) of the table that remained after pruning the major diagonal from the initial table. Once the nonlinear ODA has been performed, the table is pruned a second time to change the value of the explained cells to zero. As seen, once the nonlinear model is removed, the remaining non-zero elements in the table are uniformly distributed, and no tertiary ODA model is possible.

Initial Table					Table After the First Pruning				
	Class Variable					Class Variable			
Attribute	1	2	3	4	Attribute	1	2	3	4
1	25	10	5	5	1	0	10	5	5
2	10	25	5	5	2	10	0	5	5
3	5	5	25	10	3	5	5	0	10
4	5	5	10	25	4	5	5	10	0

On what basis does one make the decision to stop this procedure—when is it inappropriate to continue this procedure for another iteration? The answer depends on one's objectives. In applications such as gene sequencing or data compression, the objective is to achieve perfect classification of all of the events in the table, using a minimum number of assignment rules to accomplish this task. Explicit identification of a solution that obtains perfect classification while ensuring a minimum number of steps (models) necessitates a MultiODA analysis. However, the procedure described here is an heuristic approach that can be used in an effort to obtain an acceptable or satisfying solution for this problem (cf. Simon, 1956). If one is interested in such an heuristic, the iterative ODA procedure is continued until all of the events in the table are correctly classified. In other applications, such as are commonly seen in applied scientific research, the objective is to ascertain the presence and nature of the systematic (nonrandom) structure underlying an empirical table. In this circumstance, the user must decide on an appropriate stopping rule. Several intuitive stopping rules for an iterative ODA analysis, some of which have parallels in traditional multiattribute statistical methodologies, are described below.

Stopping Rules for Iterative Analyses

Experimentwise Type I Error

Using this stopping rule, retain only models that are statistically significant at the experimentwise criterion (Armitage, McPherson, & Rowe, 1969; Yarnold, 1992).

Absolute Gain in Overall Percentage Accuracy in Classification (PAC) (or Standardized Effect Strength Index)

Using this stopping rule, retain only models for which the associated absolute gain in overall PAC—computed as the number of tabled events correctly classified by the model divided by the total number of events in the original table (or an absolute gain measure based on a standardized effect index)—is at least as great as some user-specified criterion. For example, retain the model only if it has an associated absolute gain of at least 5% in overall PAC.

Relative Gain in Current PAC

Using this stopping rule, retain a model only if its associated relative gain in current overall PAC (or standardized effect strength index) relative to—that is, "over and above"—the model derived in the prior analysis (computed as the number of tabled events correctly classified by the model at the current step divided by the number of tabled events correctly classified by the model at the prior step) is at least as great as some user-specified criterion. For example, retain the model only if it is associated with a relative gain of 50% or more in overall PAC relative to the model identified in the preceding step.

Cumulative Overall PAC

Using this stopping rule, retain all of the ODA models—including the model identified in the current step—that have been identified at the first occasion that their associated cumulative overall PAC—that is, the number of tabled events correctly classified by the system of ODA models divided by the total number of events in the original table (or a comparable measure using a standardized effect index) is equal to or greater than some user-specified criterion. For example, retain ODA models identified via iterative decomposition only if, and as soon as, together they obtain a cumulative overall PAC of at least 75%.

Number of Steps (or Models)

Using this stopping rule, retain all ODA models identified in the first s steps of the iterative procedure. For example, retain all ODA models identified in the first three steps of an iterative decomposition.

Parsimony

All other things being equal, if theory A (i.e., the set of ODA models identified in the iterative analysis) explains the same phenomena as well or nearly as well as theory B, but theory A has fewer facets, assumptions, or other "moving parts" than theory B, then theory A is preferred over theory B on the basis of parsimony. For example, if theory A correctly classifies 85% of the tabled events using four ODA models, and theory B correctly classifies 87% of the events using five ODA models, theory A is preferred over theory B on the basis of parsimony.

Relative Efficiency

This stopping rule contrasts two theories on the basis of their relative "bang for the buck." That is, if theory A correctly classifies 80% of the events in the transition table using four ODA models (corresponding to a mean of 20% correct classification per model), and theory B correctly classifies 90% of the events in the transition table using six ODA models (a mean of 15% correct classification per model), then theory A would be preferred over theory B on the basis of having greater relative efficiency.

Interpretability

This stopping rule favors solutions (collections of ODA models) that are highly face valid, consistent with a priori theory, or otherwise more interpretable than alternative solutions. For example, imagine that a five-model solution allowed correct classification of 65% of the events in a transition table (a mean of 13% correct classification per model), and that a seven-model solution allowed correct classification of 91% of the transition events (a mean of 13% correct classification per model). The five- and seven-model solutions have equivalent relative efficiencies, and thus cannot be discriminated from each other on that basis. However, imagine that the structure identified by the five-model solution was highly consistent with substantive theory (and a priori hypotheses), whereas the structure identified by the seven-model solution was less interpretable in light of substantive theory. Here, the five-model solution would be preferred over the seven-model solution on the basis of its greater interpretability in the context of established theory.

Structural Decomposition With Sequential Data

We use four increasingly complex substantive examples as a means of illustrating structural decomposition with sequential data. In the first example we investigate structure underlying carbonate units, performing structural decomposition of a Markov transition table. In the second and third examples we explore the temporal stability of learning styles, performing structural decompositions of turnover tables. In the second example consecutive codes may repeat (data may fall into the major diagonal), and in the third example consecutive codes may not repeat (the major diagonal is null). Finally, in the fourth example we investigate longitudinal voting intentions in the context of a multisample analysis.

Example 12.1

To illustrate nondirectional ODA-based iterative structural decomposition of a Markov state transition table, we return to Example 11.1, involving investigation of structure underlying carbonate units. We first establish the following a priori three-component stopping rule. Continue iteration and retain all ODA models identified at each step until one of the following rules is violated: (a) the ODA models are statistically significant at experimentwise $p < .05$; (b) each step is associated with an absolute gain of at least 15% in overall PAC; and (c) each step has an associated relative efficiency—defined as the ratio of cumulative overall PAC divided by the number of steps—of at least 20%-per-step. Note that, because the associated relative efficiency is a minimum of 20%-per-step, a maximum of five steps is theoretically possible. Considered as a whole, this stopping rule mandates that, if it is possible to do so, ODA will use a small number of statistically significant models identified in a few efficient steps to maximize model effect strength for sensitivity (the analysis is weighted by prior odds).

Step 1 was already reported in Example 11.1: A single ODA model was identified that correctly classified 217 of the total of 514 events in the transition table. The statistics required by the stopping rule are overall PAC = mean absolute gain in overall PAC = mean PAC-per-step = 42.22%; and estimated $p < 0.0001$. As these results did not violate any of the stopping rule components, the primary ODA model was retained, and we proceeded to step 2.

Step 2 began by pruning the original state transition table (ex111.dat) by changing the entries (values) in the six cells that were classified correctly in step 1 to zero, in order to create a new state transition table (ex121a.dat) containing 514 − 217 = 297 transition events. Nondirectional ODA was then performed on this new table using the same program that was used in step 1. Step 2 analysis identified two ODA models—that is, a two-dimensional solution—that correctly classified 136 of the 297 events in the new transition table. Symbolically, the substructures implied by the secondary ODA models (using the fastest moving starting heuristic) were

$$D \rightarrow B \rightarrow C$$

and

$$A \rightarrow G \rightarrow F \rightarrow E.$$

Because the three-rock sequence was associated with 15—and the four-rock sequence with 121—correctly predicted events in the transition table, the former sequence occurred at an order of magnitude slower rate than the latter sequence. This raises the possibility that the faster (four-rock) sequence dominated alternative solutions to the extent that the additional 15 correctly classified events for the slower (three-rock) sequence were superfluous, or were contrived (by ODA) so as to be able to accommodate the four-rock sequence in the model. Future research in this area should investigate degenerate decomposition models, which would not require all class categories to enter into a classification model at every step of the procedure.

Statistics required by the stopping rule are overall PAC for two-step solution = $[(217 + 136) / 514] \times 100\% = 68.68\%$; absolute gain in overall PAC at step 2 = $(136 / 514) \times 100\% = 26.46\%$; mean PAC-per-step = $68.68\% / 2 = 34.33\%$; and estimated $p < 0.0001$. Because these results did not violate any of the components of the stopping rule, both secondary ODA models identified in step 2 were retained, and stepping proceeded to step 3.

Step 3 began by pruning the state transition table created in step 2 by changing the entries in the six cells that were classified correctly in step 2 to zero, in order to create a new state transition table (ex121b.dat) containing 514 − 217 − 136 = 161 transition events. Nondirectional ODA was then performed on this new table using the same program used in the preceding steps. Step 3 analysis identified a single ODA model that together correctly classified 73 of the 161 events in the new transition table. The statistics required by the stopping rule are overall PAC for three-step solution = [(217 + 136 + 73) / 514] × 100% = 82.88%; gain in overall PAC at step 3 = (73 / 514) × 100% = 14.20%; mean PAC-per-step = 82.88% / 3 = 27.63%; and estimated $p < 0.0001$. Because the result for absolute gain in overall PAC (14.2%) was less than the criterion (15%), the tertiary ODA model was dropped and the analysis is terminated.

From the perspective of the user, the next activity would involve interpreting the ODA models identified by step 2 of the iterative decomposition in light of both a priori hypotheses (if they exist) and post hoc substantive theory. This, of course, is true so long as even one model is retained. That is, the user should attempt to identify or construct theoretical propositions that support or explain the sequential structure identified by the analysis. Whether or not theoretical support is forthcoming, it is a sound idea to attempt to replicate such exploratory findings.

Example 12.2

We return to Example 11.2, involving Loo's (1996) data on the temporal stability of learning styles. Extraction of the first ODA model—which was directional, positing that data would be stable and fall along the major diagonal—left 53.7% of the sample remaining misclassified. A second, exploratory, analysis was therefore conducted to determine whether any statistically reliable sequential structure underlay off-diagonal elements. This was accomplished by appending the following ODA script at the end of the program for Example 11.2:

```
OPEN ex122.dat;
OUTPUT ex122.out;
DIRECTIONAL OFF;
GO;
```

The resulting ODA model for the turnover table pruned of all diagonal entries was, If type at time 1 = Assimilator, then predict type at time 2 = Accommodator; otherwise, if type at time 1 = Accommodator, then predict type at time 2 = Diverger; otherwise, if type at time 1 = Converger, then predict type at time 2 = Assimilator; otherwise, if type at time 1 = Diverger, then predict type at time 2 = Converger. This model correctly classified 34 (42.50%)

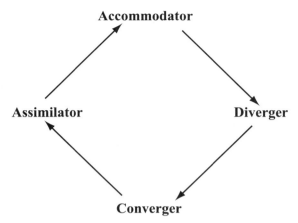

Figure 12.1. Loo's (1996) data on the temporal stability of learning styles: reliable transitions.

of the total of 80 off-diagonal entries in the pruned turnover table. This result was statistically significant (estimated $p < 0.0053$) and, although stable in LOO analysis, classification performance was of weak practical strength (overall effect strength = 24.26%). A schematic illustration of the transition profile implied by this ODA model is given in Figure 12.1. Arrows indicate the direction of transitions.

The ODA analyses presented here precisely support Loo's eyeball analysis. That eyeball analysis correctly identified the stability and change models is not surprising, because optimal classification rules are relatively obvious for small nonweighted turnover tables and for larger turnover tables (e.g., having seven or more categories) for which there are strong effects. The task becomes more difficult as the number of categories increases and as the effect strength decreases. Of course, as the effect strength weakens, the data in the rows of the turnover table tend toward a uniform distribution. In the limit, there will be no statistically significant stability or change model—a condition that also is obvious by eyeball examination.

Example 12.3

As an example of the use of ODA to assess directional nonlinear hypotheses for a turnover table in which consecutive codes may not repeat (the major diagonal is structural zeros), reconsider Foa's (1971) data on the exchange of material and psychological resources. In the

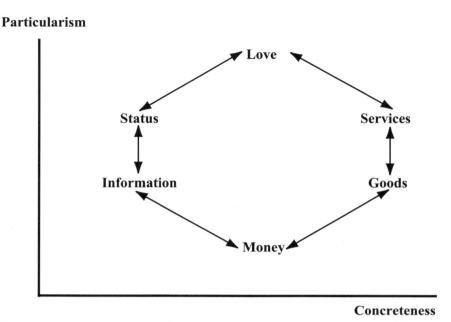

Figure 12.2. Loo's (1996) data on the temporal stability of learning styles: similar pairings.

following schematic illustration of Foa's *similarity* hypothesis, double arrows indicate the hypothesized most *similar* pairings of resource categories (see Figure 12.2).

To test the similarity hypothesis, 37 people received three messages for each resource category: Separately for each received message, a person returned a message—selected from an accompanying deck—judged to be most like the message received. In the deck all categories were represented except the category from which the message was received. The resulting turnover table is given below. Note that, although row marginals were constrained to be equivalent, column marginals were (and are) not. And, in addition to the diagonal structural zeros, matters were further complicated by three cells having five or fewer entries. Accordingly, Foa (1971) presented an eyeball analysis of the turnover table:

> With a few exceptions, the highest frequencies in each row or column are in the two cells bordering the main diagonal. Frequencies in the cells two steps removed from the diagonal are lower and the lowest frequency is in the cell which is three steps removed, and thus the most distant from the diagonal. (p. 346)

Message Received (Time 1)	Message Returned (Time 2)					
	Love	Status	Information	Money	Goods	Services
Love	0	5	21	48	29	8
Status	4	0	19	27	30	31
Information	20	11	0	20	25	35
Money	56	10	21	0	4	20
Goods	42	18	27	6	0	18
Services	12	20	37	26	16	0

Two directional alternative hypotheses are implied by Foa's similarity hypothesis: one for similarity assessed in the counter-clockwise direction and another for similarity assessed in the clockwise direction, as indicated in the schematic diagram. Consider the counter-clockwise hypothesis. To understand how to specify this hypothesis using the DIRECTIONAL command (the short form is DIR in ODA script), start with the first type of message received. As indicated in the table above, love—coded as 1—is the first type of message received; status—coded as 2—is the second type of message received; and services—coded as 6—is the sixth type of message received. From the figure, in the counter-clockwise direction, love (1) is followed by status (2) [thus the first parameter code in the DIR command is "2"]; status (2) is followed by information (3) [thus the second parameter code in DIR is "3"]; and services (6) is followed by love (1) [thus the sixth parameter code in DIR is "1"]. The following ODA script was used to conduct the counter-clockwise hypothesis.

```
OPEN ex123a.dat;
OUTPUT ex123a.out;
CATEGORICAL ON;
TABLE 7;
CLASS COL;
DIRECTIONAL < 2 3 4 5 6 1;
GO;
```

This directional hypothesis achieved an overall classification accuracy of 34.98%: The classification performance was weak in practical terms (total effect strength = 20.59%), but statistically significant (estimated $p < 0.0001$).

The second half of Foa's similarity hypothesis involves the clockwise direction. Here love (1) is followed by services (6) [thus the first DIR parameter code is "6"]; status (2) is followed by love (1) [thus the second DIR parameter code is "1"]; and services (6) is followed by goods (5) [thus the sixth DIR parameter code is "5"]. The following

ODA script was substituted into the above script to evaluate the clockwise hypothesis for the pruned table.

```
OPEN ex123b.dat;
OUTPUT ex123b.out;
DIRECTIONAL < 6 1 2 3 4 5;
GO;
```

This directional hypothesis achieved an overall classification accuracy of 51.04%. The classification performance was moderate in practical terms (total effect strength = 37.45%), but statistically significant (estimated $p < 0.0001$). Considered together, these two a priori models explained $(233 + 221) / 666 \times 100\%$, or 68.2% of the total number of events in the original table.

Example 12.4

Assessing stability and change in turnover tables is also straightforward for multisample applications. As an illustration of this, consider data provided by Goodman (1962), who used five turnover tables to report the responses of 445 people who completed six consecutive monthly interviews concerning their voting intentions of the 1940 presidential election (consecutive codes were allowed to repeat). Note that the combination of a relatively small sample and relatively strong stability resulted in sparse cells and marginal imbalance for each of the five transition tables. Nevertheless, Goodman evaluated the stationarity of the transition probabilities in a first-order Markov chain using log-linear models. Analyses suggested that the May–June and June–July tables are similar to each other and that they are different from the August–September and September–October tables, which are similar to each other. Goodman noted that the July–August table, which was different from the other four tables, reflected the time period during which the Democratic convention was held, and also that the only significant difference between the July–August table and the August–September and September–October tables involved the undecided category.

GenODA may be used to assess whether stability and change phenomena generalize across the five turnover tables (as there are only three type codes presently, a maximum of two models are possible). Considering first stability, it is hypothesized that voting intentions are primarily stable, so data will fall in the major diagonal of each turnover table. Thus, in the first step of the analysis, a single directional model—which specifies that voting intention at time $t + 1$ will equal voting intention at time t—is simultaneously imposed on each separate turnover table. Prior odds weighting is used because of class category sample size imbalance. LOO analysis is superfluous. Analysis was accomplished using the following ODA script.

```
OPEN ex124a.dat;
OUTPUT ex124a.out;
CATEGORICAL ON;
TABLE 3;
CLASS COL;
GEN TABLE 5;
DIRECTIONAL < 1 2 3;
MC ITER 10000;
GO;
```

As seen in Table 12.1, for each turnover table the stability hypothesis was statistically significant (experimentwise $p < 0.05$) and yielded total effect strength that was strong in practical terms. Clearly, the data in each table primarily reflect stable voting intentions.

To assess change, the tables were pruned by setting the diagonal elements equal to zero. Then a nondirectional analysis was conducted using the following ODA script.

```
OPEN ex124b.dat;
OUTPUT ex124b.out;
CATEGORICAL ON;
TABLE 3;
CLASS COL;
GEN TABLE 5;
MC ITER 10000;
GO;
```

Summarized in the table above, for every turnover table this model was statistically significant (experimentwise $p < 0.05$) and yielded moderate total effect strength. The resulting GenODA model is illustrated schematically in Figure 12.3 (movement occurs in a clockwise direction). As seen, the primary direction of change was from Undecided to Democrat, followed by from Republican to Undecided. Transitions clearly favored the Democratic Party in this example. Model classification performance was stable in LOO validity analysis for all turnover tables except July–August. And in the July–August table, model LOO sensitivity was stable for the Democrat and Republican classes, but degraded for the Undecided category.

Considered together, these findings suggest that all five turnover tables shared a common transition profile, but the effect was unstable for people who were undecided during the July–August time period. The latter result corroborates Goodman's (1962) findings obtained via log-linear analyses.

Table 12.1

May	June			Stability		Change	
	Republican	Democrat	Undecided	ES	p	ES	p
Republican	125	5	16	75.8	.0001	34.7	.0001
Democrat	7	106	15				
Undecided	11	18	142				

June	July			Stability		Change	
	Republican	Democrat	Undecided	ES	p	ES	p
Republican	124	3	16	76.9	.0001	23.2	.0053
Democrat	6	109	14				
Undecided	22	9	142				

July	August			Stability		Change	
	Republican	Democrat	Undecided	ES	p	ES	p
Republican	146	2	4	71.6	.0001	33.0	.0001
Democrat	6	111	4				
Undecided	40	36	96				

August	September			Stability		Change	
	Republican	Democrat	Undecided	ES	p	ES	p
Republican	184	1	7	85.1	.0001	41.4	.0001
Democrat	4	140	5				
Undecided	10	12	82				

September	October			Stability		Change	
	Republican	Democrat	Undecided	ES	p	ES	p
Republican	192	1	5	85.3	.0001	30.4	.0001
Democrat	2	146	5				
Undecided	11	12	71				

Note. Adapted from Goodman (1962). Hypotheses tested were stability (data fall in the major diagonal) and change (a nondirectional hypothesis was tested). ES = total effect strength (0 = chance; 100 = perfect classification). P is the estimated generalized permutation probability for classification accuracy achieved using the ODA model.

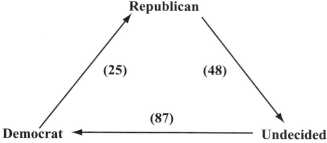

Figure 12.3. Goodman's (1962) data on voting intentions: *generalizable* transitions.

Reliability, Bias, and Random Error

According to classical test theory (CTT), any given *observed score* is constituted by the true value of the attribute in the object of measurement (*true score*) plus unmeasured sources of variability (*error score*). The fundamental equation of CTT is observed score = true score + error score. A central concept in CTT is the idea of reliability—an index that is theoretically derived on the basis of the fundamental CTT equation. Chapter 7 demonstrates how to use ODA to evaluate the reliability of a measure using a (non)weighted inter-rater, parallel form, split-half or test–retest methodology. For CTT, reliability implies stability: Testing this linear hypothesis in ODA simply involves predicting that the data will fall in the diagonal of the reliability table. However, we also noted that in the ODA paradigm, in addition to the linear model, nonlinear processes (and corresponding ODA models) may be hypothesized to underlie data in reliability tables. Of course, these nonlinear models might also be identified in exploratory analysis. Thus, these data clearly showed some stability.

In addition to these earlier concepts, the iterative decomposition methodology suggests that *both* linear and nonlinear structures may underlie the data in a reliability table. That is, first use ODA to assess stability ("reliability") in the measure—indicated by data falling in the major diagonal—and then, via iterative structural decomposition, also assess possible structure ("bias") underlying off-diagonal data. Thus, ODA can assess both reliability and bias for a measurement methodology. The CTT fundamental equation is thus amended: observed score = true score + bias score + error score (for observations in the diagonal, bias score = 0; for observations off-diagonal, true score = 0).

ODA facilitates assessment of both the statistical and practical significance of stability and bias. The null hypothesis is that there is no stability or bias. That is, the (off)diagonal responses of one (rater, split-half, parallel form, or testing) cannot be predicted by the (off)diagonal responses of the other (rater, split-half, parallel form, or testing), and the (non)directional, (non)weighted alternative hypothesis is that stability or bias exists.

Example 12.5

As an example of an iterative structural decomposition of a reliability table, we return to Example 7.1, which involved two cardiologists independently classifying each of 200 electrocardiograms. In the inter-rater reliability table, rows represented cardiologist X's classifications and columns represented cardiologist Y's classifications: For both rows and columns, 1 = normal, 2 = possibly abnormal, and 3 = definitely abnormal. In Example 7.1 a priors-weighted ODA was conducted that tested the directional alternative hypothesis that cardiologists' ratings were reliable: That is, events in the inter-rater contingency table would fall in the major diagonal. The ODA stability model achieved an overall PAC of 65.00% across the three diagnoses: Classification performance was statistically significant and of moderate practical significance. Although this analysis suggests that the two cardiologists provided consistent diagnoses of electrocardiograms, there nevertheless remains a nontrivial level of inconsistent joint ratings by the cardiologists left in the cross-classification table. Accordingly, a pruned cross-classification table was created in which the entries in the diagonal of the original table were replaced by zeros. Then, a nondirectional ODA was conducted for the pruned table in order to identify the nature and the statistical and practical significance of any systematic structure underlying the yet-unexplained (off-diagonal) data. Note that, because the stability model correctly classified 65% of the events in the original cross-classification table, the best that any subsequent model could hope to achieve would be to explain those 35% of the events in the original table that were incorrectly classified by the stability model. The following ODA script was used to conduct the nondirectional bias analysis.

```
OPEN ex125.dat;
OUTPUT ex125.out;
CATEGORICAL ON;
TABLE 3;
CLASS ROW;
LOO;
MC ITER 10000;
GO;
```

The resulting ODA bias model was, If row = 1, then column = 2 (if cardiologist X rates an electrocardiogram as normal, then cardiologist Y rates the electrocardiogram as possibly abnormal); if row = 2, then column = 3 (if X rates an electrocardiogram as possibly abnormal, then Y rates it as definitely abnormal); and finally, if row = 3, then column = 1 (if X rates an electrocardiogram as definitely abnormal, then Y rates it as normal). This model correctly classified 60 of the 70 events (85.71%) in the pruned table: Classification accuracy, stable in LOO analysis, was very strong in practical terms (total effect strength = 81.25%) and statistically significant (estimated $p < 0.0001$). Iterative ODA analysis of reliability tables may proceed until terminated by an a priori stopping rule or by the absence of any remaining misclassifications in the cross-classification table.

These results suggest that cardiologists' joint diagnoses of the electrocardiograms were primarily (65%) reliable (consistent), but also that 60 / 200 = 30% of the joint diagnoses were explained by the secondary structure identified in the ODA bias analysis. This secondary structure suggested that classifications made by cardiologist Y were one level more severe than classifications made by cardiologist X, except when the classification made by cardiologist X was "definitely abnormal," in which case the corresponding classification made by cardiologist Y was "normal." The data for this example were hypothetical, and it may be unlikely that this specific finding would actually explain the pattern of inconsistencies (i.e., nature of the bias) in joint diagnoses that occur between experts evaluating actual data. On the other hand, the latter assignment rule was associated with only 10 correct classifications, and may have been included as part of the second model by fiat—that is, in order to allow the model to successfully explain the two other cells, which—with 60 observations versus 10—are worth substantially more to the ODA optimization algorithm. Thus, future research in this area should consider degenerate decomposition models, which do not classify into all class categories in every model.

Validity, Bias, and Random Error

As in discussion regarding the conventional conceptualization of reliability, the CTT-based conceptualization of validity involves the notion that data should fall in the major diagonal of the validity contingency table for convergent validity, or outside the diagonal for discriminant validity. In this model, error is constituted by off-diagonal data for convergent validity, or by data in the diagonal for discriminant validity. In either case, it is assumed that: observed score = true score + random error. As in the discussion concerning reliability analysis, for validity analysis one may use iterative structural decomposition to determine whether table elements misclassified by the primary (convergent or discriminant) validity directional alternative hypothesis reflect random error or whether there is evidence of underlying systematic bias in the data.

Example 12.6

As an example of an iterative structural decomposition of a convergent validity table, consider Nishikawa, Kubota, and Ooi's (1983) data concerning correspondence between methods for assigning proteins into theoretical *types*. A sample of 325 proteins was assigned into one of four mutually exclusive and exhaustive types—either α, β, $\alpha + \beta$, or α / β (dummy-coded as 1 through 4, respectively)—twice, separately, on the basis of two independent theoretical methods. One method assessed *biological characteristics* of the proteins (method 1), and the other method assessed *amino acid compositions* of the proteins (method 2). Whereas both methods assign a given protein into one of four theoretical types, the question is whether they assign a given protein into the identical type: a test of convergent validity. For ODA, the priors-weighted directional alternative hypothesis is that type assignments directly correspond between methods: Data will fall in the major diagonal. The null hypothesis is that this is not true: Codes assigned to proteins by the two methods do not correspond. Data were entered in tabular form: rows were codes for biologically defined type (multicategory class variable), and columns were codes for amino-acid-composition-defined type (polychotomous attribute). The convergent validity hypothesis was tested using the following ODA script.

```
OPEN ex126a.dat;
OUTPUT ex126a.out;
CATEGORICAL ON;
TABLE 4;
CLASS ROW;
DIRECTIONAL < 1 2 3 4;
MC ITER 10000;
GO;
```

The resulting directional ODA model achieved 66.46% overall classification accuracy, which was statistically significant (estimated $p < 0.0001$), and relatively strong in practical terms (total effect strength = 51.09%). These findings suggest that there is relatively strong, statistically and practically significant convergent validity between protein-typing methods. However, because these methods were not perfectly convergent (overall PAC was less than 100%), iterative structural decomposition may be conducted to assess whether off-diagonal elements in the validity contingency table reflect a systematic assignment bias between biological versus amino-acid-composition protein-typing methods (the exploratory alternative hypothesis). The null hypothesis is that there is no such systematic bias: off-diagonal elements are uniformly distributed. Accordingly, the original validity table was pruned—the cells in the diagonal (already correctly predicted) were replaced by zeros. Then, the following ODA script was substituted into the above program and used to evaluate the bias analysis.

```
OPEN ex126b.dat;
DIRECTIONAL OFF;
LOO;
GO;
```

The resulting exploratory ODA bias model achieved overall classification accuracy of 50.46% that was statistically significant (estimated $p < 0.0001$), stable in LOO validity analysis, and of moderate strength when considered in practical terms (total effect strength = 37.10%). The structure of this exploratory model was, If type = α by amino acid method, then predict type = β by biological method; if type = β by amino acid method, then predict type = α by biological method; if type = $\alpha + \beta$ by amino acid method, then predict type = α / β by biological method; otherwise, if type = α / β by amino acid method, then predict type = $\alpha + \beta$ by biological method. This model indicates local unreliability: Between-method assignments are from one type of protein to the neighboring type of protein, and assignments do not cross over from one of the first two types to one of the second two types of proteins. Thus, even though there is strong evidence that these protein-typing methods are convergently valid, they manifest a nontrivial level of systematic inter-method classification bias. Considered together these two ODA models correctly classified (216 + 55) / 325 = 83.4% of the events in the original cross-classification table. An a priori stopping rule may be invoked at this point to limit the number of iterations conducted.

EPILOGUE

The Future of ODA

ODA constitutes a new paradigm in the statistical analysis of data. It is intuitively appealing, in the mathematical modeling of any process, that one's model make as few mistakes as possible—this is the essence of the ODA paradigm. The fruitfulness of the ODA paradigm, particularly in its application to the analysis of previously unanalyzable problems, is an indication of its value in service as a general-purpose problem-solving tool. Reflecting the synergistic amalgamation of two disciplines, the ODA paradigm joins the powerful modeling capabilities of mathematical programming and the powerful inferential capabilities of statistics. Because the ODA paradigm is inherently distribution- and metric-free, it avoids the necessity of making the simplifying assumptions required by conventional analytic techniques. And it is possible to blend the different ODA techniques so that each problem can be formulated in terms of its own unique characteristics.

What does the future hold for the ODA paradigm and for those who use and study ODA? Because the ODA paradigm is in the early stages of its development, it is obviously impossible to answer these questions with more than a modicum of certainty. However, a large and rapidly growing body of research investigates issues relevant to the ODA paradigm. In fact, since the 1970s—and especially in the last decade—such research has been published and presented at conferences with geometrically increasing frequency: The term *zeitgeist* is appropriate. This research addresses a range of issues that extends far beyond those considered in this book and, as is true of all disciplines, the majority of the most advanced research is not yet published. Such research is being conducted in our laboratory, of course, and in other laboratories (of which we are aware) in Canada, France, Ireland, England, Bulgaria, Portugal, Japan, India, Spain, Russia, the Netherlands, and the United States. This activity notwithstanding, individuals using ODA software to analyze their data are essentially at the cutting edge of this new paradigm.

To explain our optimism concerning the prospects for continued development of the ODA paradigm, we simply ask you to recall the breadth of analyses that are possible using the present ODA software. Everything this software does—*everything*—is accomplished by moving step-by-step through a single stream of data, placing markers (cutpoints) between neighboring numbers that are not the same, looking either one (if you tell it to) or all (if you don't) directions up and down the data stream, and counting the number of errors. Given the simplicity of this UniODA (univariable ODA, meaning a single attribute is used in the analysis) approach, and given its usefulness across a wide variety of applications, we marvel at its unparalleled alliance of theoretical parsimony and empirical muscle.

It is difficult to imagine, therefore, what the MultiODA (multivariable ODA—two or more attributes are used simultaneously in the analysis) paradigm is capable of, because it is allowed to wander beyond the line—to look at the entire picture rather than simply at the space between two neighboring data points. As is true for the UniODA paradigm, MultiODA identifies a model that maximizes overall (non)weighted percentage accuracy in classification (PAC) for any given problem. However, MultiODA does so by exploring hyperdimensional spaces and, at the end of its search, it identifies classification functions that are expressed in terms of multidimensional shapes—convex polyhedral cones—rather than in terms of one or more optimal cutpoints. We are working on a book for the *Optimal Data Analysis* series, specifically on MultiODA formulations (with accompanying software of course). Here we briefly describe some issues concerning MultiODA generally, as well as specific MultiODA formulations that our forthcoming book will present.

Empirically speaking, relatively little is known about the comparative performance of ODA versus suboptimal traditional statistical alternatives. Although a body of research compares the performance of MultiODA versus conventional procedures (e.g., see Erenguc & Koehler, 1990; Joachimsthaler & Stam, 1988; Rubin, 1990b; Stam & Joachimsthaler, 1989; Stam & Jones, 1990), most research involves Monte Carlo simulation and is thus difficult to generalize to real world data. Clearly, there is a need for research that compares results obtained using ODA with results obtained using traditional alternatives, *particularly in the analysis of real world data*. This need constitutes a wealth of opportunity for important, straightforward studies that should be of great interest across scientific disciplines: a true "research bonanza."

Theoretically speaking, two preeminent research areas motivate those working on the development of MultiODA as a viable statistical paradigm. First is the discovery of closed-form solutions (cf. Strube, 1991) for MultiODA distributions for any given data configuration. Because to our knowledge only two presentations (Yarnold & Soltysik, 1992a, 1992b) and one article (Yarnold, Soltysik, & Martin, 1994) discuss estimation of Type I error for Multi-ODA solutions, further development in this domain clearly constitutes a research priority.

Second is the need to obtain fast solutions to MultiODA problems. It seems hackneyed to cite the increasing speed of personal computers and access to supercomputing technology—in the context of a partial solution. However, increasingly speedy computing technology has a noticeable effect on the size of realistically solvable MultiODA problems that should not be underemphasized. Nevertheless, greatest gains in the realistic size of solvable problems can only be realized by advances in the process by which MultiODA solutions are obtained.

For example, a promising approach to solving MultiODA problems involves the Warmack–Gonzalez (1973) search technique. Using this technique, we have obtained a reduction of more than an order of magnitude in computation time as compared to recent mixed-integer mathematical programming approaches. To describe this algorithm, we first define a dichotomy as the set of inequalities that are satisfied for a particular solution. Every dichotomy is associated with a cone bounded by a frontier of one or more edges: If there are a attributes, an edge is defined as the intersection of a hyperplanes corresponding to a observations. The Warmack–Gonzalez algorithm iterates in a nonexhaustive manner through sequences of edges in the course of obtaining all optimal cones. Each edge is evaluated by solving an $a+1$ by a system of homogeneous linear equations, and then measuring the number of misclassifications obtained by the solution. The number of edges that are evaluated is always less than the total number of edges. To apply this technique it is necessary to assume the Haar condition—that is, that the observations are in general position so that it is possible to evaluate any edge. Rubin (1999) extended this methodology by relaxing the Haar condition.

Soltysik and Yarnold (1994b) used actual applications as well as Monte Carlo research to investigate the computer resources required by this algorithm as a function of the number of observations, number of attributes, and relative discriminability of the data. Problems involving two attributes and 697 observations can be solved in less than one CPU minute on an IBM 3090. This also holds true for problems with three attributes and 202 observations, and for problems with four attributes and 100 observations. Our findings suggest that the number of attributes exerts greater influence on computation time than does either the number of observations or the relative discriminability of the data.

The idea of evaluating edges gives rise to a powerful and extremely fast heuristic for finding approximate MultiODA solutions for large problems. The first step of this heuristic involves evaluating classification performance for each of a number of random edges and saving the best solution. The second step involves replacing observations one at a time from the incumbent edge with misclassified observations: This step is repeated until no further improvement is found. The final step involves repetitively solving the linear program described previously (the rows of this program are composed of the satisfied inequalities in the incumbent solution, along with one currently violated inequality): A positive optimal value for the objective function indicates an improvement in the solution. Via Monte Carlo research, we found this procedure outperformed another very fast heuristic for solving MultiODA problems (Koehler & Erenguc, 1990): For 90 problems with three random attributes and 100 observations, our heuristic obtained the optimal model in 45 problems and never missed the optimal value by more than three misclassifications.

General-Purpose MultiODA Models

A major thrust of current research involves development of comprehensive general-purpose MultiODA methods accomplished via mathematical programming. For example, Soltysik and Yarnold (1992b) developed the MIP45 mixed-integer programming formulation in which the discriminant function is normalized so that the sum of the absolute values of the coefficients is one. After selecting an optimal model (if more than one is found) using the same a priori selection heuristics used presently, a unique MultiODA solution is computed by solving a linear program. The solution to this program maximizes the minimum deviation over all satisfied inequalities, subject to the normalization previously discussed. If this deviation is positive, then the MultiODA solution associated with the system of strict inequalities has been found, and classification gaps and ambiguities in predicted class membership that exist in some other formulations do not arise. If the problem data are all integer and the discriminant coefficients are limited to a bounded set of integers, a variant of the MIP45 formulation may be used that always finds optimal solutions according with the system of strict inequalities.

The MIP45 approach can be extended to obtain MultiODA solutions that maximize the priors-, and/or cost- or return-weighted number of satisfied inequalities. Another extension of the MIP45 formulation involves fixing the sign of the discriminant coefficients (e.g., when one has a priori knowledge of the direction of an effect): In fact, bounds or any linear constraints on the coefficients may be imposed. Yet another type of constraint that can be modeled is any Boolean function of actual or predicted class membership among the observations. For example, one could force some observations to be correctly classified in the MultiODA solution (if this is feasible). Another illustration of this procedure would involve forcing observation A to be assigned to a certain class only if observation B is similarly assigned (it is notable that each of these constraints serves to increase the speed with which optimal solutions are identified).

Using MIP45, a technique for reducing the problem size can be applied when multiple observations share identical values for all attributes: Observations may be aggregated into a single observation by applying a weight to the objective function (this is especially useful if all attributes are binary). With this procedure, MultiODA problems with five binary attributes and ten million observations can routinely be solved in less than 10 CPU seconds on an IBM 3090.

When conducting MultiODA, it is desirable to obtain a solution with as few terms as possible, in light of the principle of parsimony. This can be achieved by setting an upper bound on the number of misclassifications and minimizing the number of attributes used in the solution. A related problem involves the determination of an optimal subset of attributes with exactly *a* members. This procedure is useful when the ratio of number of attributes to number of observations is too high to yield a meaningful model or when redundant or meaningless attributes are present. It should be noted that multiple optimal subsets of attributes may exist, in which case heuristic selection strategies should be used. In applied research we have found that this procedure often requires significantly greater computer resources than does standard MIP45. Thus, it may often be best to begin the analysis with a assuming small values, and to increase a until satisfactory or asymptotic classification performance is achieved.

Heretofore we have only considered the linear case. However, it is important to note that MultiODA may be easily extended to a large class of nonlinear separating surfaces. This is accomplished by defining attributes that are polynomial functions of the original attributes—any nonlinear function that is linear in the parameters of the original attributes may be modeled in this way. Of course, the resulting model is linear in its coefficients. Finally, it is also possible to solve problems involving more than two classes, in two ways. First, if there are c categories of the class variable, determine the MultiODA solution obtained with $c - 1$ separating surfaces in parallel with each other. From a computational standpoint, this is the equivalent of adding an extra attribute for each additional class category. As a second method, one may determine c different discriminant functions. An observation is then assigned to the class category for which the maximum value is obtained over these functions. Given a original attributes, from a computational standpoint this is the equivalent of a MultiODA problem with $a \times c$ attributes. A computer program that creates SAS/OR code to run the MIP45 formulation for a MultiODA problem is available (Soltysik & Yarnold, 1991).

Special-Purpose MultiODA Models

In addition to the general MultiODA model, new special-purpose MultiODA models have also been developed that warrant additional research (Soltysik & Yarnold, 1992a). Our descriptions of these new MultiODA models will be brief.

BooleanODA

The ODA approach of minimum classification error may be applied to classification problems with purely logical attributes (cf. Page, 1977). In this case, the decision rule involved in the assignment of an observation to a class is a Boolean function of logical attributes that have been measured for that observation. That is, we wish to find a Boolean function with at most t terms that minimizes the number of classification errors. Alternatively, we may look for a function with at most m misclassifications that minimizes the number of logical terms. Either way, these problems can be formulated as integer programs or solved by exhaustive enumeration. As an illustration of this method, consider an application in which two emergency physicians independently diagnosed whether 51 patients with hip trauma had a bony abnormality. Each physician rated the patients as normal or abnormal on the PPP test (a measure of sound conduction) and as normal or abnormal on visual inspection. The presence/absence of bony abnormality (class variable) was independently determined radiographically. BooleanODA identified a single optimal decision that achieved 96% overall PAC (versus 75% via logistic regression). The optimal model was, If either physician rates either attribute as abnormal, then classify the observation as abnormal; otherwise classify the observation as normal.

TauODA

Another fruitful area of investigation relates to the use of MultiODA in the analysis of data that have been sorted into ordered categories. The motivation underlying this procedure involves maximizing the goodness-of-fit between the actual and predicted category assignments. Kendall's tau is a similarity index widely used for comparison of two ranked sequences, that is proportional to the number of satisfied inequalities between paired observations. TauODA finds a linear discriminant function that maximizes the value of Kendall's tau (this differs from the multicategory case, which corresponds to the analysis of unordered categories). Illustrating this procedure, Soltysik and Yarnold (1993) considered the problem of ranking residency applicants for a program in General Internal Medicine. For each of 41 applicants in 1988, three attributes were measured: board scores, average scores obtained from ratings of letters of recommendation, and overall performance on four desirability indices scored by a committee. The consensus ranking of the applicants was obtained, and then TauODA was used to identify the optimal prediction model. An optimal Kendall's tau of .641 was obtained. The optimal model was then applied to the set of applicants for 1989, from which a tau value of .456 was obtained.

Another promising application of TauODA involves a metric-free approach to multiple linear regression. It differs from classical regression in that the criterion involved is the preservation of the rank order of the observations on the dependent measure, rather than the sum of squared errors between the values on the dependent measure and the predicted values (Yarnold, Soltysik, Curry, & Martin, 1989). This is accomplished using a linear model that maximizes Kendall's tau. Because in this case the rank order of the predicted values (rather than the magnitude of the deviations) is of primary interest, this approach should theoretically be less sensitive to the presence of outliers in the training data than would be the case in classical regression. The first step in the analysis involves sorting the observations by their values on the dependent measure. Next, a system of homogeneous strict linear inequalities between all pairs of observations is established (pairs of observations with tied values on the dependent measure have their associated inequalities dropped). TauODA is then applied to this inequality system to find the optimal predicted rank sequence. The resulting solution is then adjusted as follows. A ray in the interior of the optimal cone is found by solving a linear program which maximizes the minimum distance to the facets forming the boundary of the cone (this step guarantees that the inequalities are satisfied strictly). A univariate L1 (absolute deviation) or L2 (squared deviation) regression is then performed, to a point on the ray that optimizes either the L1 or L2 error criterion. The solution of this univariate regression problem is the optimal model.

Soltysik and Yarnold (1993) investigated the properties of this TauODA regression procedure using Monte Carlo experiments. In all, 400 experiments were performed for each of 7 models: In each experiment there were 3 attributes and 30 observations. Both training and validity data sets were generated, half of which were correlated standard normal data and the other half normal data with outliers (obtained by multiplying the value of the dependent variable by 10 for 25% of the observations). Each data set was analyzed with classical regression, TauODA regression, and L1- and L2-weighted TauODA regression. For each TauODA regression, both the L1 and L2 adjustments were performed. The problems were solved using the Warmack–Gonzalez algorithm (Soltysik and Yarnold, 1994b) run on a 50 MHz 486 microcomputer. For all problems, the values of R^2, Kendall's tau, and L1- and L2-weighted tau were recorded. The most striking result was found in the analysis of 3-way interactions with R^2 as the dependent variable. When the seven models were trained without outliers they performed similarly, with a moderate drop-off in R^2 when outliers were present in the validity set. When the models were trained with outliers, however, the L1-adjusted TauODA models outperformed the others in validity sets both with and without outliers. For the case in which outliers were present in the training set and not in the validity set, the L1-adjusted models outperformed the others by a huge margin. In addition, these models showed stable performance across all data sets trained with outliers. Thus, the TauODA regression technique with L1 adjustment is a promising alternative to classical multiple regression, especially in the case when outliers may exist in the dependent measure.

ExactODA

In some problems observations are available for which class membership is unknown. Often, exactly o of these observations are to be acted on in some manner. The initial phase of the ExactODA approach to this problem involves the partitioning of the observations into two sets: the decision set—which consists of observations with unknown class membership, and the evaluation set—which consists of observations with known class membership. To illustrate this procedure, consider the problem of selecting j job applicants from a pool of applicants. The attributes may reflect measures of previous employment experience and of skills required to perform the job task. The evaluation set is comprised of previously hired individuals who have been measured on these attributes. Each individual in the evaluation set is weighted by a performance index: here, by a measure of job performance. The decision set is comprised of the pool of job applicants who have been measured on the attributes, j of whom are to be selected for employment. ExactODA finds a solution that maximizes the weighted number of inequalities in the evaluation set such that exactly j inequalities in the decision set are satisfied. As another example, consider the problem of selecting prisoners to be released under a court mandate which requires that exactly p must be released, because of overcrowding. In this example the decision set is the current population of prisoners and the evaluation set are those prisoners who previously have been released. The performance index, which is to be minimized, is a measure of mayhem produced by the previously released prisoners. Our current research studies the application of ExactODA in the areas of market research (direct mail) and investment.

IntegerODA

We recently discovered that UniODA may be used to solve MultiODA problems in which the discriminant coefficients are constrained to take on a small set of values. For example, in a problem with a attributes; discriminant coefficients restricted to the values 0, 1, or -1; and an unconstrained threshold coefficient (i.e., cutpoint); all optimal solutions may be found by solving $(3^a / 2) - 1$ UniODA problems (Yarnold, Soltysik, Lefevre, & Martin, 1998). In general, for a problem with c possible coefficient values and a attributes, $(c^a / 2) - 1$ UniODA problems are solved. As a byproduct of this analysis, optimal attribute subsets of every size are evaluated. If c and a are relatively small, then computational problems do not arise because of the fast speed of UniODA. For example, we analyzed a problem with 900 observations, three coefficient values, and eight attributes in 716 seconds using a 33 MHz 386 microcomputer.

TemplateODA

Another interesting MultiODA application involves the design of optimal templates. In this situation, an individual is given a list of questions, along with a set of possible responses for each question, one of which is to be selected as the individual's answer to the question (e.g., each question is answered by "filling in" a circle corresponding to a selected answer on a standardized answer sheet). The actual class membership status of each observation is known. The objective of this procedure is to produce a template—that is, a series of "holes" on an opaque sheet—so that by laying the template over an individual's answer sheet and counting the number of filled-in circles, a discriminant score is produced for the individual. This score is then compared to the cutpoint obtained by the model in order to assign class membership to the individual: This assignment minimizes the number of classification errors for the training sample. In preliminary research, TemplateODA has been formulated as a pure integer problem (Soltysik & Yarnold, 1992a). For example, a 38-item questionnaire (each item answered as being "true" or "false") was completed by 107 employees of a corporation, 70 of whom were known to be desirable workers, and 37 of whom were known to be undesirable workers. For this problem, TemplateODA identified a template which resulted in 74.8% overall PAC, requiring 26 CPU minutes to solve on an IBM 3090 mainframe computer running SAS/OR.

Additional interesting directions for future research include the use of MultiODA in obtaining optimal "receiver operator characteristic" (Green & Swets, 1966; Kraemer, 1992; Loke, 1989; Swets, 1992), information theory (Kullback & Keegel, 1984; O'Neill, 1993), and tree-structured (Loh & Vanichsetakul, 1988) models, and the incorporation of ideas from fuzzy set theory (Rubin, 1992; Smithson, 1987). Clearly, it is also possible that MultiODA applications may find special utility when programmed as special-purpose computer chips. For example, ODA chips might be dedicated to the tasks of minimizing the storage and/or transmission costs of data compression, or of maximizing return-weighted decision making in automated investment arbitrage.

Nonlinear Classification Tree Analysis

A problem with any linear model—such as multiple regression analysis, logistic regression analysis, or Fisher's or Smith's discriminant analysis (Grimm & Yarnold, 1995)—involves three implicit assumptions (Yarnold, 1996b; Yarnold, Soltysik, & Bennett, 1997). First, it assumes that the attributes it uses are *the* important attributes for all observations. In contrast, nonlinear models may use different attributes to classify different segments (partitions) of the total sample. Second, it assumes that the attributes it uses have the same direction of influence—negatively or positively predictive of group membership status—for all observations. In contrast, in nonlinear models attributes may be negatively predictive of group membership status for one partition and positively predictive for another partition. Third, it assumes that the attributes it uses have the same coefficient value (decision weight) for all observations. In contrast, in nonlinear models attributes may have one coefficient value for one partition and different coefficient values for other partitions.

Several nonlinear classification methods have been developed, and preliminary research suggests they yield greater accuracy than linear models (e.g., Kass, 1980; Magidson, 1988; Wagner, Young, & Patty, 1994). Because these nonlinear methods use traditional statistical paradigms such as least squares or maximum likelihood to obtain the classification model, they do not explicitly maximize training classification accuracy. In contrast, hierarchically optimal classification tree analysis—hereafter abbreviated as CTA—involves iterative use of UniODA to construct a nonlinear multiattribute "tree" model for discriminating observations from different classes. CTA is analogous to principal components analysis, in which successive eigenvectors explain a monotonically decreasing proportion of the total variance (Grimm & Yarnold, 1995). In CTA however, successive attributes ("nodes") classify with maximum (weighted or nonweighted) accuracy a monotonically decreasing proportion of the total sample.

The first step of CTA involves conducting UniODA for each potential attribute, for the total training sample. After obtaining a UniODA model for every attribute, select the model yielding the greatest effect strength for sensitivity and compute Type I error. When constructing ("growing") the tree, attributes are retained if they have generalized (per-comparison) $p < 0.05$.

The second step of CTA involves only those observations (i.e., that partition) predicted to be members of Class 0 in the preceding step (for simplicity assume a two-category class variable). Of those observations, some are correctly classified (in reality they are members of Class 0), and some may be misclassified (in reality they are members of Class 1). It is desirable to improve classification accuracy for this partition if accuracy is imperfect. Thus, UniODA is conducted using all the potential attributes, in an effort to improve classification accuracy for this partition.

Using the ODA software, this is accomplished via the INCLUDE and EXCLUDE commands. For example, imagine the root (initial) node of one's CTA model involved the attribute age, and that the left branch involved all observations 28 years of age or younger. That is, the UniODA model predicting outcome as a function of age was, If age ≤ 28 then outcome = 0. To stratify the sample one may use the command, INCLUDE AGE<28. Or, one may specify, EXCLUDE AGE>28.

If no attribute(s) yield(s) improved mean sensitivity then this branch of the classification tree is terminated. However, if an attribute is identified that improves mean sensitivity, and that has generalized $p < 0.05$, then this new attribute is added as a node to this branch of the tree. If an additional attribute is identified, then the associated UniODA model will make two types of classification predictions: Observations will be predicted to be members of Class 0 or of Class 1 on the basis of their score on the selected attribute and the *cutpoint* or *category*. This procedure is iterated until all steps (i.e., branches of the classification tree) have been terminated.

After the tree can no longer be grown, nodes are removed ("pruned") if their Type I error exceeds the criterion given by a sequentially rejective Sidak Bonferroni-type multiple comparisons procedure. This serves as a safeguard against model overfitting and ensures that the investigator-specified experimentwise Type I error rate is maintained. Another safeguard against overfitting involves pruning nodes that show evidence of instability, as indicated in a LOO (jackknife) validity analysis.

Finally, as is true for any UniODA-based analysis, three summary indices of effect strength are particularly useful in evaluating the performance of any classification model: the sensitivity, predictive value, and total (mean of sensitivity and predictive value) effect strengths. The effect strength indices facilitate direct comparison of the classification performance achieved by models that differ with respect to sample size and imbalance, number of response categories for the class variable, and/or number and/or metric of attributes.

CTA has been used with remarkable success in a variety of areas, often providing dramatic improvement in both effect strength and substantive richness of the findings, versus traditional suboptimal methods and linear ODA. Examples of studies analyzed via CTA include discriminating geriatric versus non-geriatric ambulatory patients on the basis of functional status information (Yarnold, 1996b); predicting in-hospital mortality of patients with AIDS-related *Pneumocystis carinii* pneumonia (Arozullah et al., 2000; Yarnold, Soltysik, & Bennett, 1997), community acquired pneumonia (Arozullah et al., 2003), or who receive CPR (Yarnold, Soltysik, Lefevre, & Martin, 1998); modeling selection for discretionary treatment in intermittent claudication (Feinglass, Yarnold, Martin, & McCarthy, 1998); evaluating aggregated single-case assessments of patient functional outcomes (Yarnold, Feinglass, Martin, & McCarthy, 1997); predicting patient satisfaction with emergency medicine care (Yarnold, Michelson, Thompson, & Adams, 1998); diagnosing attention deficit disorders (Ostrander, Weinfurt, Yarnold, & August, 1998); modeling substance abuse of severely mentally ill patients (Mueser et al., 2000); predicting functional outcomes of patients with asthma (Kucera, Greenberger, Yarnold, Choy, & Levenson, 1999) or chronic fatigue syndrome (Collinge, Yarnold, & Raskin, 1998); evaluating transformational breath work and immunoenhancement (Collinge & Yarnold, 2001); predicting deep vein thrombosis in stroke patients (Green, Hartwig, Chen, Soltysik, & Yarnold, 2003); automating emergency medicine triage decision-making (Handler et al., in press); identifying different types of violent offenders and predicting violent recidivism (Stalans, Yarnold, Seng, Olson, & Repp, 2004); longitudinal modeling of progression to AIDS and death (Kanter et al., 1999); or modeling antecedents of mortal drug interactions (Zakarija et al., 2004).

Most of our applied analysis lately involves CTA that *is easily conducted using the present ODA software*. In fact, all published examples of CTA have been conducted using a beta version of the present ODA software. We presently are working on a book in this series specifically on this analysis, with accompanying software of course. Until automated CTA software is available, the present software may be used, using the IN and EX commands to stratify the sample (interested readers should begin with Yarnold, 1996b, to learn how this is accomplished).

Users of ODA

As our parting thought we hazard to propose the following hypotheses concerning the plight of ODA users (note that no specific temporal ordering is implied). First, one might consider the possible effect of the ODA paradigm on the statistically and computationally advanced individuals who become familiar with it. For example, it is likely that increasing numbers of statisticians, mathematicians, computer scientists, and operations researchers will become interested in the ODA paradigm and publish important discoveries that advance the understanding of its capabilities. It is also likely that increasing numbers of such researchers will publish empirical studies that advance the understanding of the performance of ODA-based methods, both relative to alternative traditional procedures, and in an absolute sense.

One might also consider the possible effect of the ODA paradigm on more statistically and computationally restrained individuals who become familiar with it. For example, it is likely that applied scientists from a variety of disciplines—who typically would use a relatively small subset of conventional statistical procedures in the majority of their substantive research—will become interested in ODA analogs to their favored designs and publish empirical studies that serve to both announce substantively important discoveries, and also to establish a broad empirical base by which to evaluate the utility of the paradigm. It is also likely that scientists will attempt previously untried analyses (reflecting new substantively meaningful issues) and publish theoretical and empirical studies that serve to advance both substantive- and ODA-specific knowledge.

Finally, it is possible to speculate about the effect of the ODA paradigm on the community of scientists—regardless of their level of knowledge or experience—who become familiar with it. For example, cognitive emphasis may begin to increasingly focus on how to more accurately predict the phenomenon under investigation. And, as scientists think more about classification accuracy, the level of classification accuracy that they obtain in their actual research may tend to increase. In consequence, and across a variety of scientific disciplines, variables which in reality facilitate high levels of classification accuracy—that is, which truly result in excellent prediction of the phenomenon under investigation—may be discovered more rapidly and with increasing frequency. Also, compared to traditional statistics, the parsimonious and parallel syntax of the ODA paradigm and the ODA software system may facilitate interscientist communications and reduce communication ambiguities. Finally, compared to students learning only traditional statistics, among those learning the ODA paradigm, a greater percentage may become more proficient in their knowledge of and ability to conduct statistical analyses, and might accomplish this more easily and more rapidly than would be true otherwise.

APPENDIX A

Dunn and Sidak Adjusted Per-Comparison p

Adjusted Type I Error Rate as a Function of the Number of Contrasts (Remaining) to Be Conducted, Under the Dunn and the Sidak Procedures, for Experimentwise Alphas of 0.05 and 0.01

Number of outstanding (remaining) contrasts	Experimentwise $p < .05$		Experimentwise $p < .01$	
	Dunn	Sidak	Dunn	Sidak
1	5.000^{-2}	5.000^{-2}	1.000^{-2}	1.000^{-2}
2	2.500^{-2}	2.533^{-2}	5.000^{-2}	5.013^{-2}
3	1.667^{-2}	1.696^{-2}	3.334^{-3}	3.345^{-3}
4	1.250^{-2}	1.275^{-2}	2.500^{-3}	2.510^{-3}
5	1.000^{-2}	1.021^{-2}	2.000^{-3}	2.009^{-3}
6	8.334^{-3}	8.513^{-3}	1.667^{-3}	1.674^{-3}
7	7.143^{-3}	7.301^{-3}	1.429^{-3}	1.435^{-3}
8	6.250^{-3}	6.392^{-3}	1.250^{-3}	1.256^{-3}
9	5.556^{-3}	5.684^{-3}	1.112^{-3}	1.117^{-3}
10	5.000^{-3}	5.117^{-3}	1.000^{-3}	1.005^{-3}
11	4.546^{-3}	4.653^{-3}	9.091^{-4}	9.133^{-4}
12	4.167^{-3}	4.266^{-3}	8.334^{-4}	8.372^{-4}
13	3.847^{-3}	3.938^{-3}	7.693^{-4}	7.729^{-4}
14	3.572^{-3}	3.658^{-3}	7.143^{-4}	7.177^{-4}
15	3.334^{-3}	3.414^{-3}	6.667^{-4}	6.699^{-4}
16	3.125^{-3}	3.201^{-3}	6.250^{-4}	6.280^{-4}
17	2.942^{-3}	3.013^{-3}	5.883^{-4}	5.911^{-4}
18	2.778^{-3}	2.846^{-3}	5.556^{-4}	5.583^{-4}
19	2.632^{-3}	2.696^{-3}	5.264^{-4}	5.289^{-4}
20	2.500^{-3}	2.562^{-3}	5.000^{-4}	5.024^{-4}

Number of outstanding (remaining) contrasts	Experimentwise $p < .05$		Experimentwise $p < .01$	
	Dunn	Sidak	Dunn	Sidak
21	2.381^{-3}	2.440^{-3}	4.762^{-4}	4.785^{-4}
22	2.273^{-3}	2.329^{-3}	4.546^{-4}	4.568^{-4}
23	2.174^{-3}	2.228^{-3}	4.348^{-4}	4.369^{-4}
24	2.084^{-3}	2.135^{-3}	4.167^{-4}	4.187^{-4}
25	2.000^{-3}	2.050^{-3}	4.000^{-4}	4.020^{-4}
26	1.924^{-3}	1.971^{-3}	3.847^{-4}	3.865^{-4}
27	1.852^{-3}	1.898^{-3}	3.704^{-4}	3.722^{-4}
28	1.786^{-3}	1.831^{-3}	3.572^{-4}	3.589^{-4}
29	1.725^{-3}	1.768^{-3}	3.449^{-4}	3.466^{-4}
30	1.667^{-3}	1.709^{-3}	3.334^{-4}	3.350^{-4}
31	1.613^{-3}	1.654^{-3}	3.226^{-4}	3.242^{-4}
32	1.563^{-3}	1.602^{-3}	3.125^{-4}	3.141^{-4}
33	1.516^{-3}	1.554^{-3}	3.031^{-4}	3.046^{-4}
34	1.471^{-3}	1.508^{-3}	2.942^{-4}	2.956^{-4}
35	1.429^{-3}	1.465^{-3}	2.858^{-4}	2.872^{-4}
36	1.389^{-3}	1.424^{-3}	2.778^{-4}	2.792^{-4}
37	1.352^{-3}	1.386^{-3}	2.703^{-4}	2.716^{-4}
38	1.316^{-3}	1.349^{-3}	2.632^{-4}	2.645^{-4}
39	1.283^{-3}	1.315^{-3}	2.565^{-4}	2.577^{-4}
40	1.250^{-3}	1.282^{-3}	2.500^{-4}	2.513^{-4}
41	1.220^{-3}	1.251^{-3}	2.440^{-4}	2.452^{-4}
42	1.191^{-3}	1.221^{-3}	2.382^{-4}	2.393^{-4}
43	1.163^{-3}	1.193^{-3}	2.326^{-4}	2.338^{-4}
44	1.137^{-3}	1.166^{-3}	2.273^{-4}	2.284^{-4}
45	1.112^{-3}	1.140^{-3}	2.223^{-4}	2.234^{-4}
46	1.087^{-3}	1.115^{-3}	2.174^{-4}	2.185^{-4}
47	1.064^{-3}	1.091^{-3}	2.128^{-4}	2.139^{-4}
48	1.042^{-3}	1.068^{-3}	2.084^{-4}	2.094^{-4}
49	1.021^{-3}	1.047^{-3}	2.041^{-4}	2.051^{-4}
50	1.000^{-3}	1.026^{-3}	2.000^{-4}	2.010^{-4}
51	9.804^{-4}	1.006^{-3}	1.961^{-4}	1.971^{-4}
52	9.616^{-4}	9.860^{-4}	1.924^{-4}	1.933^{-4}
53	9.434^{-4}	9.674^{-4}	1.887^{-4}	1.897^{-4}
54	9.260^{-4}	9.495^{-4}	1.852^{-4}	1.862^{-4}
55	9.091^{-4}	9.323^{-4}	1.819^{-4}	1.828^{-4}
56	8.929^{-4}	9.156^{-4}	1.786^{-4}	1.795^{-4}

Appendix A

Number of outstanding (remaining) contrasts	Experimentwise $p < .05$		Experimentwise $p < .01$	
	Dunn	Sidak	Dunn	Sidak
57	8.772^{-4}	8.996^{-4}	1.755^{-4}	1.764^{-4}
58	8.621^{-4}	8.841^{-4}	1.725^{-4}	1.733^{-4}
59	8.475^{-4}	8.691^{-4}	1.695^{-4}	1.704^{-4}
60	8.334^{-4}	8.546^{-4}	1.667^{-4}	1.675^{-4}
61	8.197^{-4}	8.406^{-4}	1.640^{-4}	1.648^{-4}
62	8.065^{-4}	8.271^{-4}	1.613^{-4}	1.621^{-4}
63	7.937^{-4}	8.139^{-4}	1.588^{-4}	1.596^{-4}
64	7.813^{-4}	8.012^{-4}	1.563^{-4}	1.571^{-4}
65	7.693^{-4}	7.889^{-4}	1.539^{-4}	1.547^{-4}
66	7.576^{-4}	7.770^{-4}	1.516^{-4}	1.523^{-4}
67	7.463^{-4}	7.653^{-4}	1.493^{-4}	1.500^{-4}
68	7.353^{-4}	7.541^{-4}	1.471^{-4}	1.478^{-4}
69	7.247^{-4}	7.432^{-4}	1.450^{-4}	1.457^{-4}
70	7.143^{-4}	7.325^{-4}	1.429^{-4}	1.436^{-4}
71	7.043^{-4}	7.222^{-4}	1.409^{-4}	1.416^{-4}
72	6.945^{-4}	7.122^{-4}	1.389^{-4}	1.396^{-4}
73	6.850^{-4}	7.025^{-4}	1.370^{-4}	1.377^{-4}
74	6.757^{-4}	6.930^{-4}	1.352^{-4}	1.359^{-4}
75	6.667^{-4}	6.837^{-4}	1.334^{-4}	1.341^{-4}
76	6.579^{-4}	6.747^{-4}	1.316^{-4}	1.323^{-4}
77	6.494^{-4}	6.660^{-4}	1.300^{-4}	1.306^{-4}
78	6.411^{-4}	6.574^{-4}	1.283^{-4}	1.289^{-4}
79	6.330^{-4}	6.491^{-4}	1.266^{-4}	1.273^{-4}
80	6.250^{-4}	6.410^{-4}	1.250^{-4}	1.257^{-4}
81	6.173^{-4}	6.331^{-4}	1.235^{-4}	1.241^{-4}
82	6.098^{-4}	6.254^{-4}	1.220^{-4}	1.226^{-4}
83	6.025^{-4}	6.179^{-4}	1.205^{-4}	1.211^{-4}
84	5.953^{-4}	6.105^{-4}	1.191^{-4}	1.197^{-4}
85	5.883^{-4}	6.033^{-4}	1.177^{-4}	1.183^{-4}
86	5.814^{-4}	5.963^{-4}	1.163^{-4}	1.169^{-4}
87	5.748^{-4}	5.895^{-4}	1.150^{-4}	1.156^{-4}
88	5.682^{-4}	5.828^{-4}	1.137^{-4}	1.143^{-4}
89	5.618^{-4}	5.762^{-4}	1.124^{-4}	1.130^{-4}
90	5.556^{-4}	5.698^{-4}	1.112^{-4}	1.117^{-4}
91	5.495^{-4}	5.636^{-4}	1.099^{-4}	1.105^{-4}
92	5.435^{-4}	5.574^{-4}	1.088^{-4}	1.093^{-4}

Number of outstanding (remaining) contrasts	Experimentwise $p < .05$		Experimentwise $p < .01$	
	Dunn	Sidak	Dunn	Sidak
93	5.377^{-4}	5.514^{-4}	1.076^{-4}	1.081^{-4}
94	5.320^{-4}	5.456^{-4}	1.064^{-4}	1.070^{-4}
95	5.264^{-4}	5.398^{-4}	1.053^{-4}	1.058^{-4}
96	5.209^{-4}	5.342^{-4}	1.042^{-4}	1.047^{-4}
97	5.155^{-4}	5.287^{-4}	1.031^{-4}	1.037^{-4}
98	5.103^{-4}	5.233^{-4}	1.021^{-4}	1.026^{-4}
99	5.051^{-4}	5.180^{-4}	1.011^{-4}	1.016^{-4}
100	5.000^{-4}	5.129^{-4}	1.000^{-4}	1.006^{-4}
101	4.951^{-4}	5.078^{-4}	9.902^{-5}	9.951^{-5}
102	4.902^{-4}	5.028^{-4}	9.805^{-5}	9.854^{-5}
103	4.855^{-4}	4.979^{-4}	9.710^{-5}	9.758^{-5}
104	4.808^{-4}	4.931^{-4}	9.616^{-5}	9.664^{-5}
105	4.762^{-4}	4.884^{-4}	9.525^{-5}	9.572^{-5}
106	4.717^{-4}	4.838^{-4}	9.435^{-5}	9.482^{-5}
107	4.673^{-4}	4.793^{-4}	9.347^{-5}	9.393^{-5}
108	4.630^{-4}	4.749^{-4}	9.260^{-5}	9.306^{-5}
109	4.588^{-4}	4.705^{-4}	9.175^{-5}	9.221^{-5}
110	4.546^{-4}	4.662^{-4}	9.092^{-5}	9.137^{-5}
111	4.505^{-4}	4.620^{-4}	9.100^{-5}	9.055^{-5}
112	4.465^{-4}	4.579^{-4}	8.930^{-5}	8.974^{-5}
113	4.425^{-4}	4.539^{-4}	8.851^{-5}	8.895^{-5}
114	4.386^{-4}	4.499^{-4}	8.773^{-5}	8.817^{-5}
115	4.348^{-4}	4.460^{-4}	8.697^{-5}	8.740^{-5}
116	4.311^{-4}	4.421^{-4}	8.622^{-5}	8.665^{-5}
117	4.274^{-4}	4.384^{-4}	8.548^{-5}	8.591^{-5}
118	4.238^{-4}	4.346^{-4}	8.476^{-5}	8.518^{-5}
119	4.202^{-4}	4.310^{-4}	8.404^{-5}	8.446^{-5}
120	4.167^{-4}	4.273^{-4}	8.334^{-5}	8.376^{-5}
121	4.133^{-4}	4.239^{-4}	8.265^{-5}	8.307^{-5}
122	4.099^{-4}	4.204^{-4}	8.198^{-5}	8.239^{-5}
123	4.066^{-4}	4.170^{-4}	8.131^{-5}	8.171^{-5}
124	4.033^{-4}	4.136^{-4}	8.065^{-5}	8.105^{-5}
125	4.000^{-4}	4.103^{-4}	8.000^{-5}	8.040^{-5}
126	3.969^{-4}	4.071^{-4}	7.937^{-5}	7.977^{-5}
127	3.937^{-4}	4.039^{-4}	7.875^{-5}	7.914^{-5}
128	3.907^{-4}	4.007^{-4}	7.813^{-5}	7.852^{-5}

Number of outstanding (remaining) contrasts	Experimentwise $p < .05$		Experimentwise $p < .01$	
	Dunn	Sidak	Dunn	Sidak
129	3.876^{-4}	3.976^{-4}	7.752^{-5}	7.791^{-5}
130	3.847^{-4}	3.945^{-4}	7.693^{-5}	7.731^{-5}
131	3.817^{-4}	3.915^{-4}	7.634^{-5}	7.672^{-5}
132	3.788^{-4}	3.886^{-4}	7.576^{-5}	7.614^{-5}
133	3.760^{-4}	3.856^{-4}	7.519^{-5}	7.557^{-5}
134	3.732^{-4}	3.828^{-4}	7.463^{-5}	7.500^{-5}
135	3.704^{-4}	3.800^{-4}	7.408^{-5}	7.445^{-5}
136	3.677^{-4}	3.771^{-4}	7.353^{-5}	7.390^{-5}
137	3.650^{-4}	3.744^{-4}	7.300^{-5}	7.336^{-5}
138	3.624^{-4}	3.717^{-4}	7.247^{-5}	7.283^{-5}
139	3.598^{-4}	3.690^{-4}	7.195^{-5}	7.231^{-5}
140	3.572^{-4}	3.664^{-4}	7.143^{-5}	7.179^{-5}
141	3.547^{-4}	3.638^{-4}	7.093^{-5}	7.128^{-5}
142	3.522^{-4}	3.612^{-4}	7.043^{-5}	7.078^{-5}
143	3.497^{-4}	3.587^{-4}	6.994^{-5}	7.029^{-5}
144	3.473^{-4}	3.562^{-4}	6.945^{-5}	6.980^{-5}
145	3.449^{-4}	3.537^{-4}	6.897^{-5}	6.932^{-5}
146	3.425^{-4}	3.513^{-4}	6.850^{-5}	6.884^{-5}
147	3.402^{-4}	3.489^{-4}	6.803^{-5}	6.837^{-5}
148	3.379^{-4}	3.466^{-4}	6.757^{-5}	6.791^{-5}
149	3.356^{-4}	3.442^{-4}	6.712^{-5}	6.746^{-5}
150	3.334^{-4}	3.420^{-4}	6.667^{-5}	6.700^{-5}
151	3.312^{-4}	3.397^{-4}	6.623^{-5}	6.656^{-5}
152	3.290^{-4}	3.375^{-4}	6.579^{-5}	6.612^{-5}
153	3.267^{-4}	3.352^{-4}	6.536^{-5}	6.569^{-5}
154	3.247^{-4}	3.331^{-4}	6.494^{-5}	6.523^{-5}
155	3.226^{-4}	3.309^{-4}	6.452^{-5}	6.484^{-5}
156	3.206^{-4}	3.288^{-4}	6.411^{-5}	6.443^{-5}
157	3.185^{-4}	3.267^{-4}	6.370^{-5}	6.402^{-5}
158	3.165^{-4}	3.246^{-4}	6.330^{-5}	6.361^{-5}
159	3.145^{-4}	3.226^{-4}	6.290^{-5}	6.321^{-5}
160	3.125^{-4}	3.206^{-4}	6.250^{-5}	6.282^{-5}
161	3.106^{-4}	3.186^{-4}	6.212^{-5}	6.243^{-5}
162	3.087^{-4}	3.166^{-4}	6.173^{-5}	6.204^{-5}
163	3.068^{-4}	3.147^{-4}	6.136^{-5}	6.166^{-5}
164	3.049^{-4}	3.128^{-4}	6.098^{-5}	6.129^{-5}

Number of outstanding (remaining) contrasts	Experimentwise $p < .05$		Experimentwise $p < .01$	
	Dunn	Sidak	Dunn	Sidak
165	3.031^{-4}	3.109^{-4}	6.061^{-5}	6.091^{-5}
166	3.013^{-4}	3.090^{-4}	6.025^{-5}	6.055^{-5}
167	2.995^{-4}	3.072^{-4}	5.989^{-5}	6.019^{-5}
168	2.977^{-4}	3.053^{-4}	5.953^{-5}	5.983^{-5}
169	2.959^{-4}	3.035^{-4}	5.918^{-5}	5.947^{-5}
170	2.942^{-4}	3.017^{-4}	5.883^{-5}	5.912^{-5}
171	2.924^{-4}	3.000^{-4}	5.849^{-5}	5.878^{-5}
172	2.907^{-4}	2.982^{-4}	5.815^{-5}	5.844^{-5}
173	2.891^{-4}	2.965^{-4}	5.781^{-5}	5.810^{-5}
174	2.874^{-4}	2.948^{-4}	5.748^{-5}	5.776^{-5}
175	2.858^{-4}	2.931^{-4}	5.715^{-5}	5.743^{-5}
176	2.841^{-4}	2.915^{-4}	5.682^{-5}	5.711^{-5}
177	2.825^{-4}	2.898^{-4}	5.650^{-5}	5.678^{-5}
178	2.809^{-4}	2.882^{-4}	5.619^{-5}	5.647^{-5}
179	2.794^{-4}	2.866^{-4}	5.587^{-5}	5.615^{-5}
180	2.778^{-4}	2.845^{-4}	5.556^{-5}	5.584^{-5}
181	2.763^{-4}	2.834^{-4}	5.525^{-5}	5.553^{-5}
182	2.748^{-4}	2.818^{-4}	5.495^{-5}	5.523^{-5}
183	2.733^{-4}	2.803^{-4}	5.465^{-5}	5.492^{-5}
184	2.718^{-4}	2.788^{-4}	5.435^{-5}	5.463^{-5}
185	2.703^{-4}	2.773^{-4}	5.406^{-5}	5.433^{-5}
186	2.689^{-4}	2.758^{-4}	5.377^{-5}	5.404^{-5}
187	2.674^{-4}	2.743^{-4}	5.348^{-5}	5.375^{-5}
188	2.660^{-4}	2.729^{-4}	5.320^{-5}	5.346^{-5}
189	2.646^{-4}	2.714^{-4}	5.292^{-5}	5.318^{-5}
190	2.632^{-4}	2.700^{-4}	5.264^{-5}	5.290^{-5}
191	2.618^{-4}	2.686^{-4}	5.236^{-5}	5.262^{-5}
192	2.605^{-4}	2.672^{-4}	5.209^{-5}	5.235^{-5}
193	2.591^{-4}	2.658^{-4}	5.182^{-5}	5.208^{-5}
194	2.578^{-4}	2.644^{-4}	5.155^{-5}	5.181^{-5}
195	2.565^{-4}	2.631^{-4}	5.129^{-5}	5.154^{-5}
196	2.552^{-4}	2.617^{-4}	5.103^{-5}	5.128^{-5}
197	2.539^{-4}	2.604^{-4}	5.077^{-5}	5.102^{-5}
198	2.526^{-4}	2.591^{-4}	5.051^{-5}	5.076^{-5}
199	2.513^{-4}	2.578^{-4}	5.026^{-5}	5.051^{-5}
200	2.500^{-4}	2.565^{-4}	5.000^{-5}	5.026^{-5}

Number of outstanding (remaining) contrasts	Experimentwise $p < .05$		Experimentwise $p < .01$	
	Dunn	Sidak	Dunn	Sidak
250	2.000^{-4}	2.052^{-4}	4.000^{-5}	4.021^{-5}
300	1.667^{-4}	1.710^{-4}	3.334^{-5}	3.351^{-5}
350	1.429^{-4}	1.466^{-4}	2.858^{-5}	2.872^{-5}
400	1.250^{-4}	1.283^{-4}	2.500^{-5}	2.513^{-5}
450	1.112^{-4}	1.140^{-4}	2.223^{-5}	2.234^{-5}
500	1.000^{-4}	1.026^{-4}	2.000^{-5}	2.011^{-5}
550	9.091^{-5}	9.327^{-5}	1.819^{-5}	1.828^{-5}
600	8.334^{-5}	8.550^{-5}	1.667^{-5}	1.676^{-5}
650	7.693^{-5}	7.892^{-5}	1.539^{-5}	1.547^{-5}
700	7.143^{-5}	7.328^{-5}	1.429^{-5}	1.436^{-5}
750	6.667^{-5}	6.840^{-5}	1.334^{-5}	1.341^{-5}
800	6.250^{-5}	6.412^{-5}	1.250^{-5}	1.257^{-5}
850	5.883^{-5}	6.035^{-5}	1.177^{-5}	1.183^{-5}
900	5.556^{-5}	5.700^{-5}	1.112^{-5}	1.117^{-5}
950	5.264^{-5}	5.400^{-5}	1.053^{-5}	1.058^{-5}
1,000	5.000^{-5}	5.130^{-5}	1.000^{-6}	1.006^{-5}

Note. The tabled adjusted alphas are presented using scientific notation. For example, for the case of 8 outstanding contrasts and an experimentwise $p < .05$, Dunn's adjusted alpha is .00625, and Sidak's adjusted alpha is is .006392. Adjusted alphas are rounded up at the fourth significant digit.

APPENDIX B

Troubleshooting: Common Problems and Their Possible Solutions

*I*t is possible that users may discover a "bug" in the ODA software. As any bugs are reported vis-à-vis the Web site (www.apa.org/books/resources/yarnoldsoltysik), of course, they will be fixed and registered users will receive announcements regarding availability of updated software. However, as we have found from our own personal experiences, when using ODA software there are a number of easy-to-make oversights that, while easily corrected, may at first induce a curious composite of enraged befuddled melancholy in both novice and experienced users. Before becoming convinced that the software is fraught with programming errors, however, it is important to check that it is indeed being used properly. Accordingly, we have tabulated the most common errors that we have committed when using this software to analyze data, and most of these errors were eventually eliminated by systematically considering the issues listed below in the context of the troublesome analyses. Thus, we recommend that users systematically assess whether any of the following oversights might be at the root of their troublesome analysis before they seek outside consulting from the Web site.

- ◆ Did you want to use the command prompt and can't find it? Click Start and Run, and command.com in Windows 98 and ME, or cmd.exe in Windows NT, 2000, or XP.
- ◆ Are you in command prompt mode and your script file name prefix is longer than 8 characters? If you are running Windows 2000 or XP, enclose the file name in quotes. Try renaming it to a shorter name in Windows 98 and ME.
- ◆ Does the system not find the ODA program? Prefix the ODA command with the path where it is located (e.g., c:\oda\oda.exe *filename*).

- Did the ODA program terminate and return to the c: prompt, failing to execute the analysis? If so, open the file you specified in the OUTPUT command (or the oda.out) file with PFE, or the editor of your choice, and check for error messages.
- Did you misspell any ODA commands, omit any required spaces or periods, or insert any additional characters such as equal signs? If the answer is yes, then spell or specify the command(s) correctly.
- Does your data file (including a HOLDOUT file, if applicable) contain any missing data that you failed to identify using the MISSING command? If the answer is yes, then use the MISSING command to identify all values that you employed to signify missing data.
- Did you forget to specify the command CATEGORICAL ON for a categorical problem? If the answer is yes, then include the CATEGORICAL ON command.
- Did you misspecify the DIRECTIONAL command? Remember that this command refers to the class variable, and not to the attribute. Also, recall that if tabular input was used, ODA dummy-codes the first row (and column) as 1, the second row (and column) as 2, and the tenth (maximum allowable) row (and column) as 10. In order to determine if your directional command is the problem, try omitting it (and only it) from your program and re-running it. If the program works without the directional command, then the directional command is likely to be the problem.
- Is your sample size too large for ODA to analyze? The largest sample size allowable is 65,535 observations. One recourse in this situation is to randomly segment the total sample into smaller samples, and use these as hold-out samples.
- After all this, is the system still acting "crazy," and failing to run the analyses that you are apparently specifying correctly? It may be that you overlooked resetting an ODA feature (e.g., CATEGORICAL, DEGEN, INCLUDE, etc.) that was appropriate for another analysis conducted earlier in the same analysis session, but which is inappropriate for the present (problematic) analysis. In order to assess if this is the case, shut down the ODA system, and then try the analysis again from the beginning.
- One of the most frequently encountered problems is that the data set is missing data points. Make sure to conduct univariate descriptive analysis to ensure data quality and database integrity.
- Are you waiting for a Monte Carlo or LOO to finish before checking to see if you specified your analysis correctly or successfully fixed a suspected error? Don't wait, hit Ctrl-Break (for CANCEL)! Before investing substantial time (even a few minutes) in an ODA analysis, first make sure that your ODA program is working properly. It's generally a good idea to specify only a few Monte Carlo runs [e.g., ITER (10)] when checking if a program is specified correctly. If, for example, Monte Carlo $p = 1$ on the basis of these ten runs, then the DIRECTIONAL command might be misspecified,

or you might have overlooked an important data transformation without which the analysis is meaningless.

◆ Did ODA fail to find a solution? First, set DIRECTIONAL to off and rerun. If a solution still cannot be found, set DEGEN to on and rerun. A last recourse might be collecting more data and rerunning analyses.

References

Alf, E., & Abrahams, N. M. (1967). Mixed group validation: A critique. *Psychological Bulletin, 67,* 443–444.

Allen, M. J., & Yen, W. M. (1979). *Introduction to measurement theory.* Monterey, CA: Brooks/Cole.

Amaldi, E. (1995). The complexity and approximability of finding maximum feasible subsystems of linear relations. *Theoretical Computer Science, 147,* 181–210.

Amaldi, E., Pfetsch, M. E., & Trotter, L. E. (2003). On the maximum feasible subsystem problem, IISs, and IIS-hypergraphs. *Mathematical Programming, 95,* 533–554.

Anderson, J. A. (1972). Separate sample logistic discrimination. *Biometrika, 59,* 19–35.

Armitage, P., McPherson, C. K., & Rowe, B. C. (1969). Repeated significance tests on accumulated data. *Journal of the Royal Statistical Society, 132,* 235–244.

Arozullah, A. M., Parada, J., Bennett, C. L., Deloria-Knoll, M., Chmiel, J. S., Phan, L., & Yarnold, P. R. (2003). A rapid staging system for predicting mortality from HIV-associated community-acquired pneumonia. *Chest, 123,* 1151–1160.

Arozullah, A. M., Yarnold, P. R., Weinstein, R. A., Nwadiaro, N., McIlraith, T. B., Chmiel, J. S., et al. (2000). A new preadmission staging system for predicting in-patient mortality from HIV-associated *Pneumocystis carinii* pneumonia in the early-HAART era. *American Journal of Respiratory and Critical Care Medicine, 161,* 1081–1086.

Arthanari, T. S., & Dodge, Y. (1981). *Mathematical programming in statistics.* New York: Wiley.

Azar, B. (1997). APA task force urges a harder look at data. *APA Monitor,* March, 26.

Bacus, J. W., & Gose, E. E. (1972). Leukocyte pattern recognition. *IEEE Transactions on Systems, Man, and Cybernetics, SMC-2,* 513–526.

Bakeman, R., & Quera, V. (1995). Log-linear approaches to lag-sequential analysis when consecutive codes may and cannot repeat. *Psychological Bulletin, 118,* 272–284.

Balogh, D. W., & Merritt, R. D. (1996). Using diagnostic efficiency statistics to evaluate the concurrent validity of the Perceptual Aberation Scale. *Journal of Personality Assessment, 66,* 321–336.

Barcikowski, R. S., & Stevens, J. P. (1975). A Monte Carlo study of the stability of canonical correlations, canonical weights and canonical variate-variable correlations. *Multivariate Behavioral Research, 10,* 353–364.

Barlow, D. H., & Hersen, M. (1984). *Single case experimental designs: Strategies for studying behavior change* (2nd ed.). New York: Pergamon.

Baumeister, R. F., & Tice, D. M. (1996). Should we abandon *p* < .05? (Editorial). *Dialogue, 11*(2), 11.

Beck, J. R., & Pauker, S. G. (1983). The Markov process in medical prognosis. *Medical Decision Making, 3,* 419–458.

Bennett, K. P., & Bredensteiner, E. J. (1997). A parametric optimization method for machine learning. *INFORMS Journal of Computing, 9,* 311–318.

Berry, K. J., & Mielke, P. W. (1985). Goodman and Kruskal's TAU-B statistic: A nonasymptotic test of significance. *Sociological Methods and Research, 13,* 543–550.

Berry, K. J., & Mielke, P. W. (1992). A family of multivariate measures of association for nominal independent variables. *Educational and Psychological Measurement, 52,* 41–55.

Billingsley, P. (1961). *Statistical inference for Markov processes.* Chicago: University of Chicago Press.

Binder, K. (1986). *Monte Carlo methods in statistical physics.* New York: Springer-Verlag.

Binder, K. (1992). *The Monte Carlo method in condensed matter physics.* New York: Springer-Verlag.

Bishop, Y. M. M., Fienberg, S. E., & Holland, P. W. (1975). *Discrete multivariate analysis.* Cambridge, England: Cambridge University Press.

Blumberg, M. S. (1957). Evaluating health screening procedures. *Operations Research, 5,* 351–360.

Blyth, C. R. (1972). On Simpson's paradox and the sure-thing principle. *Journal of the American Statistical Association, 67,* 364–366.

Boring, E. G. (1950). *History of experimental psychology* (2nd ed.). New York: Appleton-Century-Crofts.

Box, G. E. P., & Draper, N. R. (1987*). Empirical modelbuilding and response surfaces.* New York: Wiley.

Box, G. E. P., & Jenkins, G. M. (1976). *Time series analysis: Forecasting and control* (2nd ed.). San Francisco: Holden-Day.

Bradley, J. V. (1968). *Distribution-free statistical tests.* Englewood Cliffs, NJ: Prentice-Hall.

Bradley, J. V. (1978). Robustness? *British Journal of Mathematical and Statistical Psychology, 31,* 144–152.

Brennan, R. L. (1983). *Elements of generalizability theory.* Iowa City, IA: ACT Publications.

Broverman, C. (1962). Normative and ipsative measurement in psychology. *Psychological Review, 69,* 295–305.

Brown, F. G. (1983). *Principles of educational and psychological testing* (3rd ed.). New York: Holt.

Brown, W. (1910). Some experimental resullts in the correlation of mental abilities. *British Journal of Psychology, 3,* 296–322.

Brownlee, K. A. (1960). *Statistical theory and methodology in science and engineering.* New York: Wiley.

Bryant, F. B. (2000). Assessing the validity of measurement. In L. G. Grimm & P. R. Yarnold (Eds.), *Reading and understanding more multivariate statistics* (pp. 99–146). Washington, DC: American Psychological Association.

Bryant, F. B., & Yarnold, P. R. (1990). The impact of Type A behavior on subjective life quality: Bad for the heart, good for the soul? *Journal of Social Behavior and Personality, 5,* 369–404.

Bryant, F. B., & Yarnold, P. R. (1995). Principal components analysis and exploratory and confirmatory factor analysis. In L. G. Grimm & P. R. Yarnold (Eds.), *Reading and understanding multivariate statistics* (pp. 99–136). Washington, DC: American Psychological Association.

Campbell, D. T., & Fiske, D. W. (1959). Convergent and discriminant validation by the multitrait-multimethod matrix. *Psychological Bulletin, 56,* 81–105.

Carmines, E. G., & Zeller, R. A. (1979). *Reliability and validity assessment.* Beverly Hills, CA: Sage.

Carmony, L., Yarnold, P. R., & Naeymi-Rad, F. (1998). One-tailed Type I error rates for balanced two-category UniODA with a random ordered attribute. *Annals of Operations Research, 74,* 223–238.

Carter, G., Chaiken, J., & Ignall, E. (1974). *Simulation model of fire department operations.* New York: Rand.

Cassidy, R. G. (1974). *Simulation of social systems: Product or process.* Ottawa: Solicitor General Canada.

Cattell, R. B. (1952). Psychological measurement: Normative, ipsative, interactive. *Psychological Review, 51,* 292–303.

Chinneck. J. W. (2001). Fast heuristics for the maximum feasible subsystem problem. *INFORMS Journal on Computing, 13,* 210–233.

Chinneck, J. W., & Dravnieks, E. E. (1991). Locating minimal infeasible constraint sets in linear programs. *ORSA Journal on Computing, 3,* 2–28.

Choy, A. C., Yarnold, P. R., Brown, J. E., Kayaloglou, G. T., Greenberger, P. A., & Patterson, R. (1995). Virus induced erythema multiforme and Stevens-Johnson syndrome. *Allergy Proceedings, 16,* 157–161.

Churchill, G. A. (1969). *Plant location analysis: A theoretical formulation.* Ann Arbor, MI: University Microfilms.

Ciccotti, G., Frenkel, D., & McDonald, I. R. (1987). *Simulation of liquids and solids: Molecular dynamics and Monte Carlo methods in statistical mechanics.* New York: North-Holland.

Clemens, W. V. (1966). An analytical and empirical examination of some properties of ipsative measures. *Psychometric Monographs, 14.*

Cliff, N. (1995). Ordinal methods in the assessment of change. In L. M. Collins & J. L. Horn (Eds.), *Best methods for the analysis of change.* Washington, DC: American Psychological Association.

Cliff, N., & Charlin, V. (1991). Variances and covariances of Kendall's tau and their estimation. *Multivariate Behavioral Research, 26,* 693–707.

Clogg, C. C., & Shihadeh, E. S. (1994). *Statistical models for ordinal variables.* Newbury Park, CA: Sage.

Cochran, W. G. (1954). Some methods of strengthening the common χ^2 tests. *Biometrics, 10,* 417–451.

Cochran, W. G., & Cox, G. M. (1957). *Experimental designs.* New York: Wiley.

Cohen, J. (1960). A coefficient of agreement for nominal scales. *Educational and Psychological Measurement, 20,* 37–46.

Cohen, J. (1994). The earth is round ($p < .05$). *American Psychologist, 49,* 997–1003.

Collinge, W., & Yarnold, P. R. (2001). Transformational breath work in medical illness: Clinical applications and evidence of immunoenhancement. *Subtle Energies & Energy Medicine, 12,* 139–156.

Collinge, W., Yarnold, P. R., & Raskin, E. (1998). Use of mind/body self-healing practice predicts positive health transition in chronic fatigue syndrome: A controlled study. *Subtle Energies & Energy Medicine, 9,* 171–190.

Conger, A. J., & Lipshitz, R. (1973). Measures of reliability for profiles and test batteries. *Psychometrika, 38,* 411–427.

Cook, T. D., & Campbell, D. T. (1978). *Quasi-experimentation: Design and analysis issues for field settings.* Chicago: Rand McNally.

Coombs, C. H., Dawes, R. M., & Tversky, A. (1970). *Mathematical psychology: An elementary introduction.* Englewood Cliffs, NJ: Prentice-Hall.

Coplin, W. D. (1968). *Simulation in the study of politics.* Chicago: Markham.

Cowles, M., & Davis, C. (1982). On the origins of the .05 level of statistical significance. *American Psychologist, 37,* 553–558.

Critchlow, D. E., & Verducci, J. S. (1992). An omnibus test for systematic changes in judges' rankings. *Journal of Educational Statistics, 17,* 1–26.

Cromack, T. R. (1989). Measurement considerations in clinical research. In C. B. Royeen (Ed.), *Clinical research handbook: An analysis for the service professions* (pp. 47–69). Thorofare, NJ: Slack.

Cronbach, L. J., Gleser, G. C., Nanda, H., & Rajaratnam, N. (1972). *The dependability of behavioral measurements: Theory of generalizability for scores and profiles.* New York: Wiley.

Cronbach, L. J., & Meehl, P. E. (1955). Construct validity in psychological tests. *Psychological Bulletin, 52,* 281–302.

Cronbach, L. J., Rajaratnam, N., & Gleser, G. C. (1963). Theory of generalizability: A liberalization of reliability theory. *British Journal of Statistical Psychology, 15,* 137–163.

Crookall, D. (1988). *Simulation—Gaming in education and training.* New York: Pergamon.

Csaki, C. (1985). *Simulation and systems analysis in agriculture.* New York: Elsevier.

Cunningham, W. H., Cunningham, I. C. M., & Green, R. T. (1977). The ipsative process to reduce response set bias. *Public Opinion Quarterly, 41,* 379–384.

Cureton, E. E. (1957). Recipe for a cookbook. *Psychological Bulletin, 54,* 494–497.

Darlington, R. B., & Stauffer, G. F. (1966a). A method for choosing a cutting point on a test. *Journal of Applied Psychology, 50,* 229–231.

Darlington, R. B., & Stauffer, G. F. (1966b). Use and evaluation of discrete test information in decision making. *Journal of Applied Psychology, 50,* 125–129.

Davenport, E. C., & El-Sanhurry, N. A. (1991). Phi/phimax: Review and synthesis. *Educational and Psychological Measurement, 51,* 821–828.

Davies, M., & Fleiss, J. L. (1982). Measuring agreement for multinomial data. *Biometrics, 38,* 1047–1051.

Dawes, R. M. (1962). A note on base rates and psychometric efficiency. *Journal of Consulting Psychology, 26,* 422–424.

Dawes, R. M. (1979). The robust beauty of improper linear models in decision making. *American Psychologist, 34,* 571–582.

Dawes, R. M., & Meehl, P. E. (1966). Mixed group validation: A method for determining the validity of diagnostic signs without using criterion groups. *Psychological Bulletin, 66,* 63–67.

DeArmon, J. S., & Lacher, A. R. (1997). Aggregate flow directives as a ground delay strategy: Concept analysis using discrete-event simulation. *Air Traffic Control Quarterly, 4,* 307–323.

de Cani, J. S. (1984). Balancing Type I risk and loss of power in ordered Bonferroni procedures. *Journal of Educational Psychology, 6,* 1035–1037.

de Wit, C. T., & Goudriaan, J. (1978). *Simulation of ecological processes.* Wageningen, The Netherlands: Centre for Agricultural Publishing and Documentation.

Disney, R. L. (1971). Probability and stochastic processes. In H. B. Maynard (Ed.), *Industrial engineering handbook* (pp. 10.32–10.51). New York: McGraw-Hill.

Dowdney, L., Rogers, C., & Dunn, G. (1993). Influences upon attendance at out patient facilities—the contribution of linear-logistic modeling. *Psychological Medicine, 23,* 195–201.

Driese, S. G., & Dott, R. H. (1984). Model for sandstone-carbonate "cyclothems" based on Upper Member of Morgan Formation (Middle Pennsylvanian) of northern Utah and Colorado. *The American Association of Petroleum Geologists Bulletin, 68,* 574–597.

Duncan, O. D., Ohlin, L. E., Reiss, A. J., & Stanton, H. R. (1953). Formal devices for making selection decisions. *American Journal of Sociology, 58,* 573–584.

Dunn, O. J. (1961). Multiple comparisons among means. *Journal of the American Statistical Association, 56,* 52–64.

Dunn, O. J., & Vardy, P. D. (1966). Probabilities of correct classification in discriminant analysis. *Biometrics, 22.*

Ebel, R. L. (1979). *Essentials of educational measurement.* Englewood Cliffs, NJ: Prentice-Hall.

Efron, B. (1975). The efficiency of logistic regression compared to normal discriminant analysis. *Journal of the American Statistical Association, 70,* 892–898.

Efron, B., & Gong, G. (1983). A leisurely look at the bootstrap, the jackknife, and cross-validation. *The American Statistician, 37,* 36–48.

Efron, B., & Tibshirani, R. (1986). Bootstrap methods for standard errors, confidence intervals, and other measures of statistical accuracy. *Statistical Science, 1,* 54–77.

Eisenbeis, R. A. (1977). Pitfalls in the application of discriminant analysis in business, finance, and economics. *The Journal of Finance, 32,* 875–900.

Eisenbeis, R. A., & Avery, R. B. (1972). *Discriminant analysis and classification procedures: Theory and applications.* Lexington, MA: Heath.

Emerson, J. D., & Moses, L. E. (1985). A note on the Wilcoxon-Mann-Whitney test for $2 \times k$ ordered tables. *Biometrics, 41,* 303.

Erenguc, S. S., & Koehler, G. J. (1990). Survey of mathematical programming models and experimental results for linear discriminant analysis. *Managerial and Decision Economics, 11,* 215–226.

Faraone, S. V., & Hurtig, R. R. (1985). An examination of social skill, verbal productivity, and Gottman's model of interaction using observational methods and sequential analysis. *Behavioral Assessment, 7,* 349–366.

Feinglass, J., Yarnold, P. R., Martin, G. J., & McCarthy, W. J. (1998). A classification tree analysis of discretionary treatment for intermittent claudication. *Medical Care, 36,* 740–747.

Feingold, M. (1992). The equivalence of Cohen's kappa and Pearson's chi-square statistics in the 2 x 2 table. *Educational and Psychological Measurement, 52,* 57–61.

Feinstein, A. R. (1988). Statistical significance versus clinical importance. *Quality of Life and Cardiovascular Care, 4,* 99–102.

Finn, M. A., & Stalans, L. J. (1996). Police referrals to shelter and mental health treatment: Examining their decisions in domestic assault cases. *Crime and Delinquency, 41,* 467–480.

Finn, S. E. (1982). Base rates, utilities, and *DSM–III:* Shortcomings of fixed-rule systems of psychodiagnosis. *Journal of Abnormal Psychology, 91,* 294–302.

Fisher, R. A. (1921). On the "probable error" of a coefficient of correlation deduced from a small sample. *Metron, 1,* 1–32.

Fisher, R. A. (1925). *Statistical methods for research workers.* Edinburgh, Scotland: Oliver & Boyd.

Fisher, R. A. (1953). *The design of experiments* (6th ed.). New York: Hafner.

Fleiss, J. L. (1986). *The design and analysis of clinical experiments.* New York: Wiley.

Fleiss, J. L., & Cohen, J. (1973). The equivalence of weighted kappa and the intraclass correlation coefficient as measures of reliability. *Educational and Psychological Measurement, 33,* 613–619.

Flexser, A. J., & Tulving, E. (1993). Recognition-failure constraints and the average maximum. *Psychological Review, 100,* 149–153.

Foa, U. (1971). Interpersonal and economic resources. *Science, 171,* 345–351.

Frank, R. E., Massy, W. F., & Morrison, G. D. (1965). Bias in multiple discriminant analysis. *Journal of Marketing Research, 2,* 250–258.

Freed, N., & Glover, F. (1986). Evaluating alternative linear programming models to solve the two-group discriminant problem. *Decision Sciences, 17,* 151–162.

Friedman, G. D. (1987). *Primer of epidemiology* (3rd ed.). New York: McGraw-Hill.

Fukunaga, K., & Kessell, D. L. (1971). Estimation of classification error. *IEEE Transactions on Computers, 20,* 1521–1527.

Gardner, W., Hartmann, D. P., & Mitchell, C. (1982). The effects of serial dependence on the use of χ^2 for analyzing sequential data in dyadic relationships. *Behavioral Assessment, 4,* 75–82.

Geisser, S. (1975). The predictive sample Reuse method with applications. *Journal of the American Statistical Association, 70,* 320–328.

Ghiselli, E. E. (1964). *Theory of psychological measurement.* New York: McGraw-Hill.

Gilbert, N. (1993). *Analyzing tabular data: Loglinear and logistic models for social researchers.* London: University of London College Press.

Gilboa, E. (1980). *Simulation of conflict and conflict resolution in the Middle East.* Jerusalem: Magnes.

Glass, G. V., Willson, V. L., & Gottman, J. M. (1975). *Design and analysis of time-series experiments.* Boulder, CO: Associated Universities Press.

Gleeson, J., & Ryan, J. (1990). Identifying minimally infeasible subsystems of inequalities. *ORSA Journal on Computing, 2,* 61–63.

Golden, R. R., & Meehl, P. E. (1979). Detection of the schizoid taxon with MMPI indicators. *Journal of Abnormal Psychology, 88,* 217–233.

Gonzalez, R., & Nelson, T. O. (1996). Measuring ordinal association in situations that contain tied scores. *Psychological Bulletin, 119,* 159.

Goodman, L. A. (1962). Statistical methods for analyzing processes of change. *American Journal of Sociology, 68,* 57–78.

Goodman, L. A. (1968). The analysis of cross-classified data: Independence, quasi-independence, and interaction in contingency tables with or without missing cells. *Journal of the American Statistical Association, 63,* 1091–1131.

Goodman, S. N., & Royall, R. (1988). Evidence and scientific research. *American Journal of Public Health, 78,* 1568–1574.

Gorsuch, R. L. (1983). Three methods for analyzing limited time-series (*N* of 1) data. *Behavioral Assessment, 5,* 141–154.

Grammer, L. C., Shaughnessy, M. A., & Yarnold, P. R. (1996). Risk factors for immunologically mediated disease in workers with respiratory symptoms when exposed to hexahydrophthalic anhydride. *Journal of Laboratory and Clinical Medicine, 127,* 443–447.

Green, D., Hartwig, D., Chen, D., Soltysik, R. C., & Yarnold, P. R. (2003). Spinal cord injury risk assessment for thromboembolism (SPIRATE Study). *American Journal of Physical Medicine and Rehabilitation.*

Green, D. M., & Swets, J. A. (1966). *Signal detection theory and psychophysics.* New York: Wiley.

Green, M. A. (1988). Evaluating the discriminatory power of a multiple regression model. *Statistics in Medicine, 7,* 519–524.

Greenblatt, R. L., Mozdzierz, G. J., Murphy, T. J., & Trimakas, K. (1992). A comparison of nonadjusted and bootstrapped methods: Bootstrapped diagnosis might be worth the trouble. *Educational and Psychological Measurement, 52,* 181–187.

Grimm, L. G., & Yarnold, P. R. (1995). *Reading and understanding multivariate statistics.* Washington, DC: American Psychological Association.

Grimm, L. G., & Yarnold, P. R. (2000). *Reading and understanding more multivariate statistics.* Washington, DC: American Psychological Association.

Grissom, R. J. (1994). Statistical analysis of ordinal categorical status after therapies. *Journal of Consulting and Clinical Psychology, 62,* 281–284.

Grove, W. M. (1985). Bootstrapping diagnosis using Baye's Theorem: It's not worth the trouble. *Journal of Clinical and Consulting Psychology, 53,* 261–263.

Guetzkow, H. S. (1963). *Simulation in international relations.* Englewood Cliffs, NJ: Prentice-Hall.

Gulliksen, H. (1950). *Theory of mental tests.* New York: Wiley.

Guttman, L. (1945). A basis for analyzing test-retest reliability. *Psychometrika, 10,* 255–282.

Guyatt, G., Walter, S., & Norman, G. (1987). Measuring change over time: Assessing the usefulness of evaluative instruments. *Journal of Chronic Diseases, 40,* 171–178.

Gynnerstedt, G., Carlsson, A., & Westerlund, B. (1977). *A model for the Monte Carlo simulation of traffic flow along two-lane single-carriageway rural roads.* Linkoping, Sweden: Statens vag och Trasikinstitut.

Haberman, S. J. (1979). *Analysis of qualitative data, Volume 2: New developments.* New York: Academic Press.

Hagen, R. L. (1997). In praise of the null hypothesis significance test. *American Psychologist, 52,* 15–24.

Hagenaars, J. A. (1990). *Categorical longitudinal data.* Newbury Park, CA: Sage.

Halperin, M., Blackwelder, W. C., & Verter, J. I. (1972). Estimation of the multivariate logistic risk function: A comparison of the discriminant function and maximum likelihood approaches. *Journal of Chronic Diseases, 24,* 125–158.

Hammesfahr, R. D. (1982). *A simulation model for the analysis of railway intermodal terminal operations.* Ann Arbor, MI: University Microfilms.

Hand, D. J. (1983). A comparison of two methods of discriminant analysis applied to binary data. *Biometrics, 39,* 683–694.

Handler, J. A., Feied, C. F., Yarnold, P. R., Sundaram, J., Gillam, M. T., Soltysik, R. C., et al. (in press). A computerized algorithm for predicting admission at the time of emergency department triage. *Annals of Emergency Medicine.*

Hanneman, R. A. (1988). *Computer-assisted theory building: Modeling dynamic social systems.* Newbury Park, CA: Sage.

Harris, C. W. (1953). Relations among factors of raw, deviation, and double-centered score matrices. *Journal of Experimental Education, 22,* 53–58.

Harris, M. M., & Schaubroeck, J. (1990). Confirmatory modeling in organizational behavior/human resource management: Issues and applications. *Journal of Management, 16,* 337–360.

Hartmann, D. P., Gottman, J. M., Jones, R. R., Gardner, W., Kazdin, A. E., & Vaught, R. S. (1980). Interrupted time-series analysis and its application to behavioral data. *Journal of Applied Behavior Analysis, 13,* 543–559.

Harvey, R. L., Roth, E. J., Yarnold, P. R., Durham, J. R., & Green, D. (1996). Deep vein thrombosis in stroke: The use of plasma d-Dimer level as a screening test in the rehabilitation setting. *Stroke, 27,* 1516–1520.

Hastad, J. (2001). Some optimal inapproximability results. *Journal of ACM, 48,* 798–859.

Hawley, D. J., & Wolfe, F. (1991). Pain, disability, and pain/disability relationships in seven rheumatic disorders: A study of 1,522 patients. *The Journal of Rheumatology, 18,* 1552–1557.

Hays, W. L. (1973). *Statistics for the social sciences* (2nd ed.). New York: Holt.

Hicks, C. R. (1973). *Fundamental concepts in the design of experiments* (2nd ed.). New York: Holt.

Hicks, L. E. (1970). Some properties of ipsative, normative, and forced-choice normative measures. *Psychological Bulletin, 74,* 167–184.

Hills, M. (1966). Allocation rules and their error rates. *Journal of the Royal Statistical Society, 28,* 1–20.

Hinkley, D. (1983). Jackknife methods. In S. Krotz (Ed.), *Encyclopedia of statistical sciences, Volume 4* (pp. 280–287). New York: Wiley.

Hintzman, D. L. (1980). Simpson's paradox and the analysis of memory retrieval. *Psychological Review, 87,* 398–410.

Hintzman, D. L. (1993). On variability, Simpson's paradox, and the relation between recognition and recall: Reply to Tulving and Flexser. *Psychological Review, 100,* 143–148.

Hochberg, Y., & Tamhane, A. C. (1987). *Multiple comparison procedures.* New York: Wiley.

Hodder, I. (1978). *Simulation studies in archaeology.* New York: Cambridge.

Holland, B. S., & Copenhaver, M. D. (1987). An improved sequentially rejective Bonferroni test procedure. *Biometrics, 43,* 417–423.

Holland, B. S., & Copenhaver, M. D. (1988). Improved Bonferroni-type multiple testing procedures. *Psychological Bulletin, 104,* 145–149.

Holm, S. (1979). A simple sequentially rejective multiple test procedure. *Scandinavian Journal of Statistics, 6,* 65–70.

Holroyd, K. A., Nash, J. M., Pingel, J. D., Cordingley, G. E., & Jerome, A. (1991). A comparison of pharmacological (amitriptyline HCL) and non-pharmacological (cognitive-behavioral) therapies for chronic tension headaches. *Journal of Consulting and Clinical Psychology, 59,* 387–393.

Hosmer, D. W., & Lemeshow, S. (1980). Goodness of fit tests for the multiple logistic regression model. *Communications in Statistics: Theoretical Methods, A9,* 1043–1069.

Hsu, L. M. (1990). Implications of inflexibility in the size of one sample for the statistical power of tests of mean contrasts. *The Journal of Applied Behavioral Science, 26,* 151–155.

Hu, L., Bentler, P. M., & Kano, Y. (1992). Can test statistics in covariance structure analysis be trusted? *Psychological Bulletin, 112,* 351–362.

Huitema, B. E. (1985). Autocorrelation in applied behavior analysis: A myth. *Behavioral Assessment, 7,* 107–118.

Hyde J. S., & Plant, E. A. (1995). Magnitude of psychological gender differences: Another side to the story. *American Psychologist, 50,* 159–161.

Ibaraki, T., & Muroga, S. (1970). Adaptive linear classifier by linear programming. *IEEE Transactions on Systems Science and Cybernetics, SSC6,* 53–62.

Isaac, P. D., & Poor, D. D. S. (1974). On the determination of appropriate dimensionality in data with error. *Psychometrika, 39,* 91–109.

Jaccard, J., Becker, M. A., & Wood, G. (1984). Pairwise multiple comparison procedures: A review. *Psychological Bulletin, 96,* 589–596.

Jackson, D. J., & Alwin, D. F. (1980). The factor analysis of ipsative measures. *Sociological Methods and Research, 9,* 218–238.

Jenkins, C. D., Zyzanski, S. J., Ryan, T. J., Flessas, A., & Tannenbaum, S. I. (1977). Social insecurity and coronary-prone Type A responses as identifiers of severe atherosclerosis. *Journal of Consulting and Clinical Psychology, 45,* 1060–1067.

Jenkins, G. M., & Watts, D. G. (1968). *Spectral analysis and its applications.* San Francisco: Holden-Day.

Joachimsthaler, E. A., & Stam, A. (1988). Four approaches to the classification problem in discriminant analysis: An experimental study. *Decision Sciences, 19,* 322–333.

Joachimsthaler, E. A., & Stam, A. (1990). Mathematical programming approaches for the classification problem in two-group discriminant analysis. *Multivariate Behavioral Research, 25,* 427–454.

Kahneman, D., Slovic, P., & Tversky, A. (1982). *Judgement under uncertainty: Heuristics and biases.* Cambridge, England: Cambridge University Press.

Kalos, M. H. (1986). *Monte Carlo methods.* New York: Wiley.

Kass, G. (1980). An exploratory technique for investigating large quantities of categorical data. *Applied Statistics, 29,* 119–127.

Kazdin, A. E. (1992). *Research design in clinical psychology* (2nd ed.). Boston: Allyn & Bacon.

Kemeny, J. G., & Snell, J. L. (1976). *Finite Markov chains.* New York: Springer-Verlag.

Kempthorne, O. (1952). *The design and analysis of experiments.* New York: Wiley.

Kendall, M. G. (1938). A new measure of rank correlation. *Biometrika, 30,* 81–93.

Keppel, G. (1982). *Design and analysis: A researcher's handbook* (2nd ed.). Englewood Cliffs, NJ: Prentice-Hall.

Keselman, H. J., Keselman, J. C., & Games, P. A. (1991). Maximum familywise Type I error rate: The least significant difference, Newman-Keuls, and other multiple comparison procedures. *Psychological Bulletin, 110,* 155–161.

King, A. C., Kiernan, M., Oman, R., Kraemer, H. C., Hull, M., & Ahn, D. (1997). Can we identify who will adhere to long-term physical activity? Signal detection methodology as a potential aid to clinical decision making. *Health Psychology, 16,* 380–389.

Kirk, R. E. (1982). *Experimental design: Procedures for the behavioral sciences* (2nd ed.). Pacific Grove, CA: Brooks/Cole.

Kleijnen, J. P. C. (1974). *Statistical techniques in simulation.* New York: Dekker.

Kleinbaum, D. G., Kupper, L. L., & Muller, K. E. (1988). *Applied regression analysis and other multivariable methods* (2nd ed.). Boston: PWS-Kent.

Klockars, A. J., Hancock, G. R., & McAweeney, M. J. (1995). Power of unweighted and weighted versions of simultaneous and sequential multiple-comparison procedures. *Psychological Bulletin, 118,* 300–307.

Koehler, G. J., & Erenguc, S. S. (1990). Minimizing misclassifications in linear discriminant analysis. *Decision Sciences, 21,* 63–85.

Kohlas, J. (1972). *Monte Carlo simulation in operations research.* New York: Springer-Verlag.

Kolb, D. A. (1984). *Experience as the source of learning and development.* Englewood Cliffs, NJ: Prentice Hall.

Kolesar, P., & Showers, J. L. (1985). A robust credit screening model using categorical data. *Management Science, 31,* 123–133.

Kramer, C. Y. (1972). *A first course in methods of multivariate analysis.* Blacksburg, VA: Kramer.

Kraemer, H. C. (1992). *Evaluating medical tests.* Newbury Park, CA: Sage.

Kruskal, J. B. (1983). An overview of sequence comparison: Time warps, string edits, and macromolecules. *SIAM Review, 25,* 201–237.

Kshirsagar, A. M. (1972). *Multivariate analysis.* New York: Dekker.

Kucera, C. M., Greenberger, P. A., Yarnold, P. R., Choy, A. C., & Levenson, T. (1999). An attempted prospective testing of an asthma severity index and a quality of life survey for 1 year in ambulatory patients with asthma. *Allergy and Asthma Proceedings, 20,* 29–38.

Kuder, G. F., & Richardson, M. W. (1937). The theory of the estimation of test reliability. *Psychometrika, 2,* 151–160.

Kullback, S., & Keegel, J. C. (1984). Categorical data problems using information theoretic approach. In P. R. Krishnaiah & P. K. Sen (Eds.), *Handbook of statistics: Vol. 4. Nonparametric methods.* New York: North-Holland.

Kussner, U., & Tidhar, D. (2000). Combining different translation sources, in Natural Language Processing – NLP 2000. Second International Conference, Patras, Greece, June 2000. In D. N. Christodoulakis (Ed.), *Lecture Notes in Computer Science 1835* (pp. 261–271). New York: Springer-Verlag.

Lachenbruch, P. A. (1967). An almost unbiased method of obtaining confidence intervals for the probability of misclassification in discriminant analysis. *Biometrics, 23,* 639–645.

Lachenbruch, P. A. (1975). *Discriminant analysis.* New York: Hafner.

Lachenbruch, P. A., & Mickey, M. R. (1968). Estimation of error rates in discriminant analysis. *Technometrics, 10,* 157–163.

Lamiell, J. T. (1981). Toward an ideothetic psychology of personality. *American Psychologist, 36,* 276–289.

Landis, J. R., & Koch, G. G. (1977). The measurement of observer agreement for categorical data. *Biometrics, 33,* 159–174.

Larichev, O. I., Olson, D. L., Moshkovich, H. M., & Mechitov, A. J. (1995). Numerical vs. cardinal measurements in multiattribute decision making: How exact is enough? *Organizational Behavior and Human Decision Processes, 64,* 9–21.

Lefevre, F., Feinglass, J., Potts, S., Soglin, L, Yarnold, P. R., Martin, G. J., & Webster, J. R. (1992). Iatrogenesis in high-risk, elderly patients. *Archives of Internal Medicine, 152,* 2074–2080.

Lefevre, F., Feinglass, J., Yarnold, P. R., Martin, G. J., & Webster, J. R. (1993). Use of the *RAND* implicit review instrument for quality of care assessment. *American Journal of the Medical Sciences, 305,* 222–228.

Lehman, R. S. (1977). *Computer simulation and modeling: An introduction.* Hillsdale, NJ: Erlbaum.

Levenson, T., Grammer, L. C., Yarnold, P. R., & Patterson, R. (1997). Cost-effective management of malignant potentially fatal asthma. *Allergy and Asthma Proceedings, 18,* 73–78.

Lilly, J. C. (1975). *Simulations of God: The science of belief.* New York: Simon & Schuster.

Loh, W. Y., & Vanichsetakul, N. (1988). Tree-structured classification via generalized discriminant analysis. *Journal of the American Statistical Association, 83,* 715–728.

Loke, W. H. (1989). Diagnostic evaluations using signal detection analysis. *Indian Journal of Psychological Medicine, 12,* 87–91.

Loo, R. (1996). Construct validity and classification stability of the revised Learning Style Inventory (LSI-1985). *Educational and Psychological Measurement, 56,* 529–536.

Lord, F. M. & Novick, M. R. (1968). *Statistical theories of mental test scores.* Reading, MA: Addison-Wesley.

Luce, R. D., & Raiffa, H. (1957). *Games and decisions: Introduction and critical survey.* New York: Wiley.

Lux, I. (1991). *Monte Carlo particle transport methods: Neutron and photon calculations.* Boca Raton, FL: CRC Press.

Lyerly, R. (1958). The Kuder-Richardson formula 21 as a split-half coefficient, and some remarks on its basic assumption. *Psychometrika, 23,* 267–270.

Maggio, R. A. (1971). Design of production and distribution systems. In H. B. Maynard (Ed.), *Industrial engineering handbook* (pp. 13.3–13.17). New York: McGraw-Hill.

Magidson, J. (1988). Improved statistical techniques for response modeling: Progression beyond regression. *Journal of Direct Marketing, 2,* 6–18.

Magnusson, D. (1967). *Test theory.* Reading, MA: Addison-Wesley.

Mangasarian, O. L. (1994). Misclassification minimization. *Journal of Global Optimization, 5,* 309–323.

Mann, H. B., & Whitney, D. R. (1947). On a test of whether one of two variables is stochastically larger than the other. *Annals of Mathematical Statistics, 18,* 50–60.

Marcotte, P., & Savard, G. (1995). A new implicit enumeration scheme for the discriminant analysis problem. *Computers and Operations Research, 22,* 625–639.

Markowski, C. A. (1990). On the balancing of error rates for LP discriminant methods. *Managerial and Decision Economics, 11,* 235–241.

Marsh, H. W., Balla, J. R., & McDonald, R. P. (1988). Goodness-of-fit indexes in confirmatory factor analysis: The effect of sample size. *Psychological Bulletin, 103,* 391–410.

Martin, E. (1981). Simpson's paradox revisited: A reply to Hintzman. *Psychological Review, 88,* 372–374.

Martin, G. J., Magid, N. M., Myers, G., Barnett, P. S., Schaad, J. W., Weiss, J. S., et al. (1987). Heart rate variability and sudden cardiac death secondary to coronary artery disease during ambulatory electrocardiographic monitoring. *American Journal of Cardiology, 60,* 86–89.

Mattavelli, M., & Amaldi, E. (1995). Using perceptronlike algorithms for the analysis and parameterization of object motion. In F. Girosi (Ed.), *Neural Network for Signal Processing V, Proceedings of the 1995 IEEE Workshop* (pp. 303–312). Cambridge, MA: IEEE Signal Processing Society.

Matthews, K. A., Krantz, D. S., Dembroski, T. M., & MacDougall, J. M. (1982). Unique and common variance in Structured Interview and Jenkins Activity Survey measures of the Type A behavior pattern. *Journal of Personality and Social Psychology, 42,* 303–313.

Mausner, B., Mausner, J. S., & Rial, W. Y. (1967). The influence of a physician on the smoking behavior of his patients. In S. V. Zagona (Ed.), *Studies and issues in smoking behavior* (pp. 103–106). Tucson: University of Arizona Press.

Maxwell, S. E., & Delaney, H. D. (1990). *Designing experiments and analyzing data: A model comparison perspective.* Belmont, CA: Wadsworth.

Maxwell, S. E., & Delaney, H. D. (1993). Bivariate median splits and spurious statistical significance. *Psychological Bulletin, 113,* 181–190.

McCleary, R., Hay, R. A., Meidinger, E. E., & McDowall, D. (1980). *Applied time series analysis for the social sciences.* Beverly Hills, CA: Sage.

McClish, D. K. (1992). Combining and comparing area estimates across studies or strata. *Medical Decision Making, 12,* 274–279.

McDowall, D., McCleary, R., Meidinger, E. E., & Hay, R. A. (1980). *Interrupted time series analysis.* Beverly Hills, CA: Sage.

McLachlan, G. J. (1992). *Discriminant analysis and statistical pattern recognition.* New York: Wiley.

Meehl, P. E. (1954). *Clinical versus statistical prediction: A theoretical analysis and a review of the evidence.* Minneapolis: University of Minnesota Press.

Meehl, P. E., & Rosen, A. (1955). Antecedent probability and the efficiency of psychometric signs, patterns, or cutting scores. *Psychological Bulletin, 52,* 194–216.

Melaragno, M. I., Smith, M. A. C., Kormann-Bortolotto, M. H., & Neto, J. T. (1991). Lymphocyte proliferation and sister chromatid exchange in Alzheimer's disease. *Gerontology, 37,* 293–298.

Mendenhall, W., & Reinmuth, J. E. (1974). *Statistics for management and economics* (2nd ed.). North Scituate, MA: Duxbury.

Mendoza, J. L., Markos, V. H., & Gonter, R. (1978). A new perspective on sequential testing procedures in canonical analysis: A Monte Carlo evaluation. *Multivariate Behavioral Research, 13,* 371–382.

Midgette, A. S., Stukel, T. A., & Littenberg, B. (1993). A meta-analytic method for summarizing diagnostic test performances: Receiver-operating-characteristic-summary point estimates. *Medical Decision Making, 13,* 253–257.

Mielke, P. W. (1984). Meteorological applications of permutation techniques based on distance functions. In P. R. Krishnaiah & P. K. Sen (Eds.), *Handbook of statistics: Vol. 4. Nonparametric methods.* New York: North-Holland.

Mielke, P. W. (1991). The application of multivariate permutation methods based on distance functions in the earth sciences. *Earth-Science Reviews, 31,* 55–71.

Mikhailor, G. A. (1992). *Optimization of weighted Monte Carlo methods.* New York: Springer-Verlag.

Modell, H. I. (1986). *Simulations in physiology: The respiratory system.* Seattle, WA: NRCLSE.

Mosteller, F. (1968). Association and estimation in contingency tables. *Journal of the American Statistical Association, 63,* 1–28.

Mosteller, F., & Tukey, J. W. (1977). *Data analysis and regression.* New York: Addison-Wesley.

Mueser, K. T., Sayres, S., Schooler, N. R., Mance, R., & Haas, G. (1993). *The reliability of the Scale for the Assessment of Negative Symptoms: A multi-cite investigation.* Manuscript submitted for publication.

Mueser, K. T., Yarnold, P. R., & Bellak, A. S. (1992). Diagnostic and demographic correlates of substance abuse in schizophrenia and major affective disorder. *Acta Psychiatrica Scandinavica, 85,* 48–55.

Mueser, K. T., Yarnold, P. R., & Foy, D. W. (1991). Statistical analysis for single-case designs: Evaluating outcomes of imaginal exposure treatment of chronic PTSD. *Behavior Modification, 15,* 134–155.

Mueser, K. T., Yarnold, P. R., Levinson, D. F., Singh, H., Bellak, A. S., Kee, K., et al. (1990). Prevalence of substance abuse in schizophrenia: Demographic and clinical correlates. *Schizophrenia Bulletin, 16,* 31–56.

Mueser, K. T., Yarnold, P. R., Rosenberg, S. D., Drake, R. E., Swett, C., Miles, K. M., et. al. (2000). Substance use disorder in hospitalized severely mentally ill psychiatric patients: Prevalence, correlates, and subgroups. *Schizophrenia Bulletin, 26,* 179–193.

Newmark, J. (1983). *Statistics and probability in modern life* (3rd ed.). Philadelphia: Saunders.

Nishikawa, K., Kubota, Y., & Ooi, T. (1983). Classification of proteins into groups based on amino acid composition and other characters, II: Grouping into four types. *Journal of Biochemistry, 94,* 997–1007.

Nishith, P., Hearst, D. E., Mueser, K. T., & Foa, E. B. (1995). PTSD and major depression: Methodological and treatment considerations in a single-case design. *Behavior Therapy, 26,* 319–335.

Noreen, E. W. (1989). *Computer-intensive methods for testing hypotheses: An introduction.* New York: Wiley.

Novick, M. R., & Lewis, G. (1967). Coefficient alpha and the reliability of composite measurements. *Psychometrika, 32,* 1–13.

Nunnally, J. C. (1964). *Educational measurement and evaluation.* New York: McGraw-Hill.

Nunnally, J. C. (1978*). Psychometric theory* (2nd ed.). New York: McGraw-Hill.

O'Neill, W. D. (1993, May). *The Kullback-Leibler distance as an optimal discriminant for predicting stock market efficiency.* Invited address presented at the TIMS/ORSA Joint National Meetings, Chicago.

Osgood, C. E., Suci, G. J., & Tannenbaum, P. H. (1957). *The measurement of meaning.* Urbana–Champaign, IL: University Press.

Ostrander, R., Weinfurt, K. P., Yarnold, P. R., & August, G. J. (1998). Diagnosing attention deficit disorders using the BASC and the CBCL: Test and construct validity analyses using optimal discriminant classification trees. *Journal of Consulting and Clinical Psychology, 66,* 660–672.

Page, C. V. (1977). Heuristics for signature table analysis as a pattern recognition technique. *IEEE Transactions on Systems, Man, and Cybernetics, SMC-7,* 77–86.

Palarine, J. R. (1972). *SOS: A simulation on simulation.* Chicago: Northwestern University.

Parker, M., & Ryan, J. (1996). Finding the minimum weight IIS cover of an infeasible system of linear inequalities. *Annals of Mathematics and Artificial Intelligence, 17,* 107–126.

Parshall, C. G., & Kromrey, J. D. (1996). Tests of independence in contingency tables with small samples: A comparison of statistical power. *Educational and Psychological Measurement, 56,* 26–44.

Parzen, E. (1962). *Stochastic processes.* San Francisco: Holden-Day.

Pavur, R. (1992, April). *Evaluating the MIP approach with a secondary goal for the classification problem.* Invited address presented at the TIMS/ORSA Joint National Meetings, Orlando, FL.

Pearson, K. (1900). On the criterion that a given system of deviations from the probable in the case of a correlated system of variables is such that it can be reasonably supposed to have arisen from random sampling. *Philosophical Magazine, 50,* 157–175.

Peters, H. J. (1985). *Investigations in landing process of aircraft by means of the Monte Carlo method.* Paris: European Space Agency.

Pindyck, R. S., & Rubinfeld, D. L. (1976). *Econometric models and economic forecasts.* New York: McGraw-Hill.

Posner, K. L., Sampson, P. D., Caplan, R. A., Ward, R. J., & Cheney, F. W. (1990). Measuring inter-rater reliability among multiple raters: An example of methods for nominal data. *Statistics in Medicine, 9,* 1103–1115.

Press, W. H., Flannery, B. P., Teukolsky, S. A., & Vetterling, W. T. (1989). *Numerical recipes: The art of scientific computing.* Cambridge, MA: University Press.

Preuss, L., & Vorkauf, H. (1997). The knowledge content of statistical data. *Psychometrika, 62,* 133–161.

Pritchett, C. W., & Roehrig, S. F. (1985). *Search and rescue Monte Carlo simulation.* Washington, DC: U.S. Department of Transportation.

Rae, G. (1991). Another look at the reliability of a profile. *Educational and Psychological Measurement, 51,* 89–93.

Rappaport, N. B., McAnulty, D. P., & Brantley, P. J. (1988). Exploration of the Type A behavior pattern in chronic headache sufferers. *Journal of Consulting and Clinical Psychology, 56,* 621–623.

Rasmussen, J. L., & Dunlap, W. P. (1991). Dealing with nonnormal data: Parametric analysis of transformed data vs. nonparametric analysis. *Educational and Psychological Measurement, 51,* 809–820.

Rau, J. G. (1970). *Optimization and probability in systems engineering.* New York: Van Nostrand.

Raush, H. L. (1965). Interaction sequences. *Journal of Personality and Social Psychology, 2,* 487–499.

Reynolds, H. T. (1977). *The analysis of cross-classifications.* New York: Free Press.

Rimm, D. (1963). Cost efficiency and test prediction. *Journal of Consulting Psychology, 27,* 89–91.

Rogers, J. H., & Widiger, T. A. (1989). Comparing ideothetic, ipsative, and normative indices of consistency. *Journal of Personality, 57,* 847–869.

Rorer, L. G., & Dawes, R. M. (1982). A base-rate bootstrap. *Journal of Consulting and Clinical Psychology, 50,* 419–425.

Rorer, L. G., Hoffman, P. J., & Hsieh, K. C. (1966). Utilities as base-rate multipliers in the determination of optimum cutting scores for the discrimination of groups of unequal size and variance. *Journal of Applied Psychology, 50,* 364–368.

Rorer, L. G., Hoffman, P. J., LaForge, G. E., & Hsieh, K. C. (1966). Optimum cutting scores to discriminate groups of unequal size and variance. *Journal of Applied Psychology, 50,* 153–164.

Rosen, A. (1954). Detection of suicidal patients: An example of some limitations in the prediction of infrequent events. *Journal of Consulting Psychology, 18,* 397–403.

Rosenthal, R. (1978). How often are our numbers wrong? *American Psychologist, 33,* 1005–1008.

Rosenthal, R. (1984). *Meta-analytic procedures for social research.* Beverly Hills, CA: Sage.

Rosenthal, R., & Rubin, D. B. (1984). Multiple contrasts and ordered Bonferroni procedures. *Journal of Educational Psychology, 6,* 1028–1034.

Rosenthal, R., & Rubin, D. B. (1985). Statistical analysis: Summarizing evidence versus establishing fact. *Psychological Bulletin, 97,* 527–529.

Rosner, B. (1982). *Fundamentals of biostatistics.* Boston: Duxbury.

Rossi, F., Sassano, A., & Smriglio, S. (2001). Models and algorithms for terrestrial digital broadcasting. *Annals of Operations Research, 22,* 625–639.

Rothke, S. E., Friedman, A. F., Dahlstrom, W. G., Greene, R. L., Arredondo, R., & Mann, A. W. (1994). MMPI-2 normative data for the FK index: Implications for clinical, neuropsychological, and forensic practice. *Assessment, 1,* 1–15.

Royeen, C. B. (1989). *Clinical research handbook: An analysis for the service professions.* Thorofare, NJ: Slack.

Rubin, P. A. (1990a). Heuristic solution procedures for a mixed-integer programming discriminant model. *Managerial and Decision Economics, 11,* 255–266.

Rubin, P. A. (1990b). A comparison of linear programming and parametric approaches to the two-group discriminant problem. *Decision Sciences, 21,* 373–386.

Rubin, P. A. (1992, April). *Mathematical programming and alternative classification models.* Invited address presented at the TIMS/ORSA Joint National Meetings, Orlando, FL.

Rubin, P. A. (1999). Adapting the Warmack–Gonzalez algorithm to handle discrete data. *European Journal of Operational Research, 113,* 632–642.

Rubinstein, R. Y. (1981). *Simulation and the Monte Carlo method.* New York: Wiley.

Rubinstein, R. Y. (1986). *Monte Carlo optimization, simulation, and queueing networks.* New York: Wiley.

Rulon, P. J. (1939). A simplified procedure for determining the reliability of a test by split-halves. *Harvard Education Review, 9,* 99–103.

Russell, C. J., & Bobko, P. (1992). Moderated regression analysis and Likert scales: Too coarse for comfort. *Journal of Applied Psychology, 77,* 336–342.

Russell, R. L., Bryant, F. B., & Estrada, A. U. (1997). Child participation in high versus low quality therapy sessions: An application of optimal discriminant analysis with P–technique. Manuscript submitted for publication.

Ryan, T. A. (1959). Multiple comparisons in psychological research. *Psychological Bulletin, 56,* 26–47.

Ryan, T. A. (1985). "Ensemble-adjusted" p values: How are they to be weighted? *Psychological Bulletin, 97,* 521–526.

Saal, F. E., Downey, R. G., & Lahey, M. A. (1980). Rating the ratings: Assessing the psychometric quality of rating data. *Psychological Bulletin, 88,* 413–428.

Sadegh, P. (1999, July). *A maximum feasible subset algorithm with application to radiation therapy.* American Control Conference, San Diego, CA.

Santow, G. (1978). *A simulation approach to the study of human fertility.* Boston: Nijhoff.

Saville, P., & Sik, G. (1991). Ipsative scaling: A comedy of measures, as you Likert or much ado about nothing. *Guidance and Assessment Review, 7,* 1–3.

Saville, P., & Willson, E. (1991). The reliability and validity of normative and ipsative approaches in measurement of personality. *Journal of Occupational Psychology, 64,* 219–238.

Scheffe, H. (1959). *The analysis of variance.* New York: Wiley.

Schiffman, S. S., Reynolds, M. L., & Young, F. W. (1981). *Introduction to multidimensional scaling: Theory, methods, and applications.* New York: Academic Press.

Schlundt, D. G., & Donahoe, C. P. (1983). Testing for independence between pairs of autocorrelated binomial data sequences. *Journal of Behavioral Assessment, 5,* 309–316.

Seaman, M. A., Levin, J. R., & Serlin, R. C. (1991). New developments in pairwise multiple comparisons: Some powerful and practicable procedures. *Psychological Bulletin, 110,* 577–586.

Shaffer, J. P. (1986). Modified sequentially rejective multiple test procedures. *Journal of the American Statistical Association, 81,* 826–831.

Shautsukova, L. Z. (1975). Solution of a classification problem by the method of calculation of estimates. *USSR Computational Mathematics and Mathematical Physics, 15,* 200–209.

Shea, J. A., Norcini, J. J., Baranowski, R. A., Langdon, L. O., & Popp, R. L. (1992). A comparison of video and print formats in the assessment of skill in interpreting cardiovascular motion studies. *Evaluation in the Health Professions, 15,* 325–340.

Shechter, M., & Lucas, R. C. (1978). *Simulation of recreational use for park and wilderness management.* Baltimore: Johns Hopkins University Press.

Sidak, Z. (1967). Rectangular confidence regions for the means of multivariate normal distributions. *Journal of the American Statistical Association, 62,* 626–633.

Simon, H. A. (1956). Rational choice and the structure of the environment. *Psychological Review, 63,* 129–138.

Simpson, E. H. (1951). The interpretation of interaction in contingency tables. *Journal of the Royal Statistical Society, B, 13,* 238–241.

Smithson, M. (1987). *Fuzzy set analysis for behavioral and social sciences.* New York: Springer-Verlag.

Snook, S. C., & Gorsuch, R. L. (1989). Component analysis versus common factor analysis: A Monte Carlo study. *Psychological Bulletin, 106,* 148–154.

Snyder, D. K., Wills, R. M., & Grady-Fletcher, A. (1991). Long-term effectiveness of behavioral versus insight-oriented marital therapy: A four-year follow up study. *Journal of Consulting and Clinical Psychology, 59,* 138–146.

Soeken, K. L., & Prescott, P. A. (1986). Issues in the use of Kappa to estimate reliability. *Medical Care, 24,* 733–741.

Sokal, R. R., & Sneath, P. H. A. (1963). *Principles of numerical taxonomy.* San Francisco: Freeman.

Soltysik, R. C., & Yarnold, P. R. (1991). MAKE45: A user interface for multivariable optimal discriminant analysis via SAS/OR. *Applied Psychological Measurement, 15,* 170.

Soltysik, R. C., & Yarnold, P. R. (1992a, April). *Special purpose optimal discriminant analyses.* Invited address presented at the TIMS/ORSA Joint National Meetings, Orlando, FL.

Soltysik, R. C., & Yarnold, P. R. (1992b, April). *Fast solutions to optimal discriminant analysis problems.* Invited address presented at the TIMS/ORSA Joint National Meetings, Orlando, FL.

Soltysik, R. C., & Yarnold, P. R. (1993). *Optimal discrimination with an ordered class variable.* Invited address presented at the TIMS/ORSA Joint National Meetings, Chicago.

Soltysik, R. C., & Yarnold, P. R. (1994a). Univariable optimal discriminant analysis: One-tailed hypotheses. *Educational and Psychological Measurement, 54,* 646–653.

Soltysik, R. C., & Yarnold, P. R. (1994b, May). The Warmack–Gonzalez algorithm for linear two-category multivariable optimal discriminant analysis. *Computers and Operations Research, 21,* 735–745.

Sorum, M. (1972). Three probabilities of misclassification. *Technometrics, 14,* 309–316.

Spearman, C. (1904). The proof and measurement of association between two things. *American Journal of Psychology, 15,* 72–101.

Spearman, C. (1910). Correlation calculated from faulty data. *British Journal of Psychology, 3,* 271–295.

Stalans, L. J., & Finn, M. A. (1995). How novice and experienced officers interpret wife assaults: Normative and efficiency frames. *Law and Society Review, 29,* 301–335.

Stalans, L. J., Yarnold, P. R., Seng, M., Olson, D. E., & Repp, M. (2004). Identifying three types of violent offenders and predicting violent recidivism while on probation: A classification tree analysis. *Law and Human Behavior, 28,* 253–271.

Stam, A., & Joachimsthaler, E. A. (1989). Solving the classification problem in discriminant analysis via linear and nonlinear programming methods. *Decision Sciences, 20,* 285–293.

Stam, A., & Jones, D. G. (1990). Classification performance of mathematical programming techniques in discriminant analysis: Results for small and medium sample sizes. *Managerial and Decision Economics, 11,* 243–253.

Sternberg, S. (1966). High-speed scanning in human memory. *Science, 153,* 652–654.

Stevens, J. (1992). *Applied multivariate statistics for the social sciences* (2nd ed.). Hillsdale, NJ: Erlbaum.

Stine, R. (1990). An introduction to bootstrap methods: Examples and ideas. In J. Fox & J. S. Long (Eds.), *Modern methods of data analysis* (pp. 325–373). Newbury Park, CA: Sage.

Stone, M. (1974). Cross-validatory choice and assessment of statistical problems. *Journal of the Royal Statistical Society, 36,* 111–147.

Strube, M. J. (1991). Multiple determinants and effect size: A more general method of discourse. *Journal of Personality and Social Psychology, 61,* 1024–1027.

Strube, M. J. (2000). Reliability and generalizability theory. In L. G. Grimm & P. R. Yarnold (Eds.), *Reading and understanding more multivariate statistics* (pp. 23–66). Washington, DC: American Psychological Association.

Student. (1908). The probable error of a mean. *Biometrika, 1,* 1–25.

Sullivan, J., & Newkirk, R. (1989). *Simulation in emergency management and technology.* San Diego, CA: SCS International.

Swets, J. A. (1992). The science of choosing the right decision threshold in high-stakes diagnostics. *American Psychologist, 47,* 522–532.

Switzer, F. S., Paese, P. W., & Drasgow, F. (1992). Bootstrap estimates of standard errors in validity generalization. *Journal of Applied Psychology, 77,* 123–129.

Taylor, J. L. (1972). *Simulation in the classroom.* Harmondsworth, PA: Penguin.

Tett, R. P., Bobocel, D. R., Hafer, C., Lees, M. C., Smith, C. A., & Jackson, D. N. (1992). The dimensionality of Type A behavior within a stressful work simulation. *Journal of Personality, 60,* 533–551.

Thompson, B. (1994). The pivotal role of replication in psychological research: Empirically evaluating the replicability of sample results. *Journal of Personality, 62,* 157–176.

Thompson, D. A., & Yarnold, P. R. (1995). Relating patient satisfaction to waiting time perceptions and expectations: The disconfirmation paradigm. *Academic Emergency Medicine, 2,* 1057–1062.

Thompson, D. A., Yarnold, P. R., Williams, D. R., & Adams, S. L. (1996). The effects of actual waiting time, perceived waiting time, information delivery, and expressive quality on patient satisfaction in the emergency department. *Annals of Emergency Medicine, 28,* 657–665.

Titterington, D. M., Murray, G. D., Murray, L. S., Spiegelhalter, D. J., Skene, A. M., Habbema, J. D. F., & Glepke, G. J. (1981). Comparison of discrimination techniques applied to a complex data set of head injured patients. *Journal of the Royal Statistical Society, 144,* 145–175.

Turner, G. E. (1985). *A simulation model of pedestrian movement in building circulation system components.* Ann Arbor, MI: University Microfilms.

Vale, C. D., & Maurelli, V. A. (1983). Simulating multivariate nonnormal distributions. *Psychometrika, 48,* 465–471.

Van Loon, J. N. M. (1981). Irreducibly inconsistent systems of linear inequalities. *European Journal of Operations Research, 8,* 282–288.

Vandamme, F. J. (1972). *Simulation of natural language: A first approach.* The Hague, The Netherlands: Mouton.

Velleman, P. F., & Wilkinson, L. (1993). Nominal, ordinal, interval, and ratio typologies are misleading. *The American Statistician, 47,* 65–72.

Wagner, E. E., Young, G. R., & Patty, C. E. (1994). Sequential optimum selection: A categorical alternative to discriminant analysis for conducting research in assessment. *Perceptual and Motor Skills, 79,* 375–383.

Wagner, M., Meller, J., & Elber, R. (2002). *Large-scale linear programming techniques for the design of protein folding potentials* (Tech. Rep. No. TR-2002—02). Norfolk, VA: Old Dominion University.

Wainer, H. (1991). Adjusting for differential base rates: Lord's Paradox again. *Psychological Bulletin, 109,* 147–151.

Wampold, B. E. (1988). Introduction [Special issue]. *Behavioral Assessment, 10,* 227–228.

Warmack, R. E., & Gonzalez, R. C. (1973). An algorithm for the optimal solution of linear inequalities and its application to pattern recognition. *IEEE Transactions on Computers, C22,* 1065–1075.

Weinfurt, K. P., Bryant, F. B., & Yarnold, P. R. (1994). The factor structure of the Affect Intensity Measure: In search of a measurement model. *Journal of Research in Personality, 28,* 314–331.

Weiss, S. A. (1983). *A simulation model for evaluating solutions to the court delay problem.* Evanston, IL: Northwestern University.

Westfall, P. H., & Young, S. S. (1993). *Resampling-based multiple testing: Examples and methods for p-value adjustment.* New York: Wiley.

Whitehurst, G. J., Fischel, J. E., DeBaryshe, B., Caulfield, M. B., & Falco, F. L. (1986). Analyzing sequential relations in observational data: A practical guide. *Journal of Psychopathology and Behavioral Assessment, 8,* 129–148.

Widiger, T. A. (1983). Utilities and fixed rules: Comments on Finn. *Journal of Abnormal Psychology, 92,* 495–498.

Wilk, H. B., Shapiro, S. S., & Chen, H. J. (1965). A comparative study of various tests of normality. *Journal of the American Statistical Association, 63,* 1343–1372.

Winer, B. J. (1971). *Statistical principles in experimental design* (2nd ed.). New York: McGraw-Hill.

Woolson, R. F. (1987). *Statistical methods for the analysis of biomedical data.* New York: Wiley.

Wright, R. E. (1995). Logistic regression. In L. G. Grimm & P. R. Yarnold (Eds.), *Reading and understanding multivariate statistics* (pp. 217–244). Washington, DC: American Psychological Association.

Wyte, C. D., Adams, S. L., Cabel, J. A., Pearlman, K., Yarnold, P. R., Morkin, M., et al. (1996). Prospective evaluation of emergency medicine instruction for rotating PGY1 residents. *Academic Emergency Medicine, 3,* 72–76.

Yarnold, B. M. (1990). *Refugees without refuge: Formation and failed implementation of U.S. political asylum policy in the 1980's.* Lanham, MD: University Press of America.

Yarnold, J. K. (1970). The minimum expectation of χ^2 goodness-of-fit tests and the accuracy of approximations for the null distribution. *Journal of the American Statistical Association, 65,* 864–886.

Yarnold, P. R. (1984). The reliability of a profile. *Educational and Psychological Measurement, 44,* 49–59.

Yarnold, P. R. (1987, April). *Androgyny and Type A behavior.* Paper presented at the Annual Meeting of the American Association for the Advancement of Science, Chicago.

Yarnold, P. R. (1988). Classical test theory methods for repeated-measures N=1 research designs. *Educational and Psychological Measurement, 48,* 913–919.

Yarnold, P. R. (1990). Androgyny and sex-typing as continuous, independent factors, and a glimpse of the future. *Multivariate Behavioral Research, 25,* 407–419.

Yarnold, P. R. (1992). Statistical analysis for single-case designs. In F. B. Bryant, L. Heath, E. Posavac, J. Edwards, S. Tindale, E. Henderson, & Y. Suarez-Balcazar (Eds.), *Social psychological applications to social issues: Vol. 2. Methodological issues in applied social research* (pp. 177–197). New York: Plenum.

Yarnold, P. R. (1994). Comparing the split-half reliability of androgyny and sex-typing measures. *Australian Journal of Psychology, 46,* 164–169.

Yarnold, P. R. (1996a). Characterizing and circumventing Simpson's paradox for ordered bivariate data. *Educational and Psychological Measurement, 56,* 430–442.

Yarnold, P. R. (1996b). Discriminating geriatric and non-geriatric patients using functional status information: An example of classification tree analysis via UniODA. *Educational and Psychological Measurement, 56,* 656–667.

Yarnold, P. R., & Bryant, F. B. (1988). A note on measurement issues in Type A research: Let's not throw out the baby with the bath water. *Journal of Personality Assessment, 52,* 410–419.

Yarnold, P. R., & Bryant, F. B. (1994). A measurement model for the Type A Self-Rating Inventory. *Journal of Personality Assessment, 62,* 102–115.

Yarnold, P. R., Bryant, F. B., Nightingale, S. N., & Martin, G. J. (1996). Assessing physician empathy using the Interpersonal Reactivity Index: A measurement model and cross-sectional analysis. *Psychology, Health, and Medicine, 1,* 207–221.

Yarnold, P. R., Feinglass, J., Martin, G. J., & McCarthy, W. J. (1997). Comparing three pre-processing strategies for longitudinal data for individual patients: An example in functional outcomes research. *Evaluation and the Health Professions, 22,* 254–277.

Yarnold, P. R., Hart, L. A., & Soltysik, R. C. (1994). Optimizing the classification performance of logistic regression and Fisher's discriminant analysis. *Educational and Psychological Measurement, 54,* 73–85.

Yarnold, P. R., Martin, G. J., Soltysik, R. C., & Nightingale, S. D. (1993). Androgyny predicts empathy for trainees in medicine. *Perceptual and Motor Skills, 77,* 576–578.

Yarnold, P. R., Michelson, E., Thompson, D. A., & Adams, S. L. (1998). Predicting patient satisfaction: A study of two emergency departments. *Journal of Behavioral Medicine, 21,* 545–563.

Yarnold, P. R., & Mueser, K. T. (1989). Meta-analyses of the reliability of Type A behaviour measures. *British Journal of Medical Psychology, 62,* 43–50.

Yarnold, P. R., Mueser, K. T., Grau, B. W., & Grimm, L. G. (1986). The reliability of the student version of the Jenkins Activity Survey. *Journal of Behavioral Medicine, 9,* 401–414.

Yarnold, P. R., & Soltysik, R. C. (1991a). Theoretical distributions of optima for univariate discrimination of random data. *Decision Sciences, 22,* 739–752.

Yarnold, P. R., & Soltysik, R. C. (1991b). Refining two-group multivariable classification models using univariate optimal discriminant analysis. *Decision Sciences, 22,* 1158–1164.

Yarnold, P. R., & Soltysik, R. C. (1992a). *Statistical distributions underlying optimal discriminant analysis.* Invited address presented at the TIMS/ORSA Joint National Meetings, Orlando, FL.

Yarnold, P. R., & Soltysik, R. C. (1992b, April). *Optimal discriminant analysis as an alternative to conventional statistical models.* Presented at the TIMS/ORSA Joint National Meetings, Orlando, FL.

Yarnold, P. R., Soltysik, R. C., & Bennett, C. L. (1997). Predicting in-hospital mortality of patients with AIDS-related *Pneumocystis carinii* pneumonia: An example of hierarchically optimal classification tree analysis. *Statistics in Medicine, 16,* 1451–1463.

Yarnold, P. R., Soltysik, R. C., Curry, R. H., & Martin, G. J. (1989, May). *In quest of the best: Resident selection based on application information and mixed integer programming.* Paper presented at the Annual Meeting of the Society of Behavioral Medicine, San Francisco.

Yarnold, P. R., Soltysik, R. C., Lefevre, F, & Martin, G. J. (1998). Predicting in-hospital mortality of patients receiving cardiopulmonary resuscitation: Unit-weighted MultiODA for binary data. *Statistics in Medicine, 17,* 2405–2414.

Yarnold, P. R., Soltysik, R. C., & Martin, G. J. (1994). Heart rate variability and inducibility for sudden cardiac death: An example of optimal discriminant analysis. *Statistics in Medicine, 13,* 1015–1021.

Yarnold, P. R., Soltysik, R. C., McCormick, W. C., Burns, R., Lin, E. H. B., Bush, T., et al. (1995). Application of multivariable optimal discriminant analysis in general internal medicine. *Journal of General Internal Medicine, 10,* 601–606.

Yarnold, P. R., Stille, F. C., & Martin, G. J. (1996). Cross-sectional psychometric assessment of the Functional Status Questionnaire: Use with geriatric versus nongeriatric ambulatory medical patients. *International Journal of Psychiatry in Medicine, 25,* 305–317.

Zakarija, A., Bandarenko, N., Pandey, D. K., Auerbach, A., Raisch, D., Kim, B., et al. (2004). Clopidogrel-associated thrombotic thrombocytopenic purpura (TTP): An update of pharmacovigilance efforts conducted by independent researchers, the pharmaceutical suppliers, and the Food and Drug Administration. *Stroke, 35,* 533–537.

Zegers, F. E. (1991). Coefficients for interrater agreement. *Applied Psychological Measurement, 15,* 321–333.

Zhuo, Z., Gatewood, L. C., & Ackerman, E. (1990). A Monte Carlo simulation program for coronary heart disease. In R. A. Miller (Ed.), *Proceedings of the Fourteenth Annual Symposium on Computer Applications in Medical Care.* Los Alamitos, CA: IEEE Computer Society Press.

Index

A

Accuracy, maximization of, 4–5
Analytic method
    as alternative to other statistical tests, 76
    closed form for theoretical distribution of optima for a priori, two-category continuous random data, 75
    open form to derive distribution of optima for post hoc hypothesis and ordered attribute, 75
    precision dimension in, 76
    theoretical dimension of, 76
ANOVA (analysis of variance)
    ODA analog of, 178
    randomized block design model, 178
        F statistics in, 179
        nuisance variation in, 178–179
ASCII files
    creation from SPSS data set, 52–54
    saving Excel data set as, 50–52
ATTRIBUTE (ATTR), variable list and, 29–30
Attribute(s)
    binary, 87–91
    continuous, 101–119
    in Fisher's method, 77
    ordered, 8
    ordinal, 93–101
    polychotomous, 91–93
    qualitative, 8
    values on, 9
Autocorrelation (time-series) analysis
    applications of, 198
    autocorrelation (ACF) function in, 198–199
    autocorrelation(k) (ACF[k]) in, 199–200
        correlogram of, 200
    ODA analogue for categorical data, 200
    notion of lag and, 199
    ODA assessment of serial dependency in time-series data with binary attribute, 200–203
    stationary process defined, 198–199
    stationary *vs.* non-stationary processes in, 198
    World Series baseball game, time series, lag-1, serial dependency of sequential baseball pitches during first two innings, 200–203
        autoregressive transition in, 201–202
        interrupted time-series in, 201
        ODA model of, 202
        polychotomous measurement in, 203
        stationarity of time series in, 202

B

Between-groups t test
    outlying data and, 102
    problems for, 102
    in two-category class variable and one continuous attribute design
        ODA *vs.*, 101
Binary attributes, defined, 87
Binary data
    directional (a priori, confirmatory, "one-tailed") hypothesis
        voting behavior of Republicans and Democrats, 90–91
    nondirectional (post hoc, exploratory, "two-tailed") hypothesis
        proportion of males and females for wasps *vs.* honey bees, 88–89
    vs. ordered data, 87–88
Bonferroni procedures, ordered, 83

Bonferroni-type procedures, sequentially-rejective *vs.* nonsequential alternatives, 82
BooleanODA, 233–234

C

CATEGORICAL (CAT), variable list and, 30
Chi-square analysis
   of data with binary class variable and binary attribute, 88
   limitation of, 88
   ODA *vs.*, 76
Classical test theory
   observed, true, error scores in, 122, 223
   observed score in, 223
   in parallel forms reliability, 128
   in sequential analysis
     single-case (N-of-1), 207–208
   validity in, 225
Classification accuracy index(es)
   overall, 58
   predictive value, 59–60
   sensitivity, 58–59
Classification performance
   accuracy in, 57
   ODA model for, 61–67
   predictive value index in, 59–60
   sensitivity index in, 57–58
   stability in, 68–70
Classification tree analysis (CTA)
   conducting UniODA for each potential attribute in, 238
   evaluation of performance in
     indices of effect strength in, 238–239
   nonlinear
     comparison with principal components analysis, 237–238
     observations predicted to be members of class 0, 238
   pruning of nodes in, 238
   studies analyzed via, 239
   termination process in, 238
   using present ODA software, 239
   utility of, 239
CLASS, variable list and, 30
Class variables
   category levels of, 7
   definition of, 7
   dichotomous, 8
   dummy codes for, 7
Commands, 28–37
Confusion table, 57–58
Construct validity. *See also* Validity
   for binary class variable and attribute
     Type A behavior and coronary artery disease, 148–149
Contingency table
   iterative (sequential) ODA-based decomposition of
     first (primary model) (non)directional ODA on original table, 210
     on one-dimensional linear solution and uniform random residuals, 210
     on one-dimensional nonlinear solution and uniform random residuals, 210–211, 211
     second model (non)directional ODA on pruned model, 210
     stopping rules for, 211, 212–214
     test data sets illustrations, 210–211
     on two-dimensional solution with primary linear model, secondary nonlinear model, and uniform residuals, 211
Continuous attributes
   directional hypothesis, binary class variable and continuous attribute
     heart rate variability and susceptibility to sudden cardiac death, 102
     Singer score in, 102, 103
   directional hypothesis, two-category class variable, continuous attribute, weighting by prior odds and return
     Dow Jones *January Barometer,* 104–105
   multicategory class variable and
     Alzheimer's disease correlates in people with and without Alzheimer's, 116–119
     Alzheimer's disease data
       cell proliferation potential results, 117
       ODA omnibus model, 117–118
       pairwise comparisons in, 118–119
       sister chromatic exchange correlate results, 117
     in completely randomized design, 115
     ODA analysis *vs.* ANOVA analysis, 116
     one-way ANOVA analysis of, 115–116
   one two-category class variable and one continuous attribute
     ODA between-groups *t* test, 101
   ordered, misclassification weighting for individual observations, 103–104

Convergent validity. *See also* Validity
   multicategory class variable and polychotomous attribute
     approaches to classification of proteins, 151–153
CTA. *See* Classification tree analysis (CTA)
Cutting scores, optimal, 10

D

DATA, 30–31
Data set
   copying from Excel, 48–52
   creation for ODA
     from Excel, 48–52
     missing data and, 55
     PFE in, 45–47
     from SAS, 54–55
     from SPSS, 52–54
   creation with PFE, 45–47
DEGEN (DEGENERATE), variable list and, 31
DIRECTION (DIR, DIRECTIONAL), variable list and, 31
Discriminant validity. *See also* Validity
   binary class variable and categorical ordinal attribute
     LOO validity analysis in, 150–151
     Type A behavior and migraine and tension headache, 150–151
Dunn's procedures
   Bonferroni-type
     in multiple comparisons, 81, 82
   nonsequential, 85
   sequentially rejective, 85, 86

E

Effect strength for sensitivity (ESS), 59, 64
Efficiency analysis, hypothetical example of, 60–61
Error rate classification, assessment of, 69–70
ESS (effect strength for sensitivity), 59, 64
ExactODA
   decision set in, 235, 236
   evaluation set in, 235–236
Example(s)
   aging and functional status (continuous, interactive attributes in) in ambulatory medical patients
     ODA-optimization of Fisher's linear discriminant analysis model, 159–160
   Alzheimer's disease correlates in people with/without AD
     multicategory class variable and continuous attribute, 116–119
   assessment of psychological androgyny in 68 male undergraduates
     split-half reliability with polychotomous attribute, 130–132
   auditing methods, efficiency of two different, nondirectional hypothesis
     binary class variable and ordinal (rank) attribute, 94–95
   cardiologists (two) classifying 200 electrocardiograms into three diagnoses
     inter-rater reliability analysis and ordinal attribute, 125–126
   collaborative treatment study of efficacy of neuroleptic maintenance dosage and family treatments for schizophrenia
     inter-rater reliability and ordinal attributes, 126–128
   congressional voting on 1826 Pinckney Gag rule
     multicategory class variable and polychotomous attribute, 109–111
   correspondence between two procedures for assessing Type A behavior
     parallel forms reliability with ordered attribute, 128–130
   dentists (four) ratings of state of ten patients' teeth
     intraclass correlation, 139–140
   exchange of material and psychological resources
     eyeball analysis of, 218
     similarity hypothesis for, 218, 219
     stability of turnover table for, 196–198
     tests of turnover table, 218
   fathers' and sons' occupational status from England and Denmark
     generalizability of fixed GenODA across multiple samples, 173–175
     hold-out validity analysis of multicategory analysis and polychotmomous attribute, 142–145
   gender and rheumatic disease (seven types), nondirectional hypothesis
     binary class variable and polychotomous attribute, 91–93

Example(s), *continued*
  gender effects *versus* other effects in psychology, directional hypothesis
    binary class variable and ordinal attribute, 100–101
  heart rate variability and susceptibility to sudden cardiac death, directional hypothesis
    binary class variable and continuous attribute, 102–103
  January Barometer, Dow Jones Industrial Index, directional hypothesis
    two-category class variable and continuous attribute, 104–105
  outcome (attribute) of marital therapy types (variable), nondirectional hypothesis
    binary class variable and ordinal attribute, 97–98
  outcomes of two headache remedies
    binary class variable and ordinal categorical attribute, 98–99
  patient overall satisfaction and satisfaction with physician (five specialties)
    generalizability of directional room-to-vary GenODA model across multiple samples, 175–178
  people completing six consecutive monthly interviews re: voting intentions in 1940 presidential election
    structural decomposition with sequential data, 220–223
  perceived waiting time and patient satisfaction with emergency department
    multicategory class variable and ordinal attribute, 113–115
  physician behavior, personal decision making, smoking behavior
    ODA-optimization of logistic regression model analysis, 161–163
  political affiliation status of high school students and parents
    multicategory class variable and polychotomous attribute, 111–112
  prospective study of efficacy of lecture and case-study teaching methods in residency program
    repeated measures (within subjects) analysis, 204–207
  proteins, classification into types, two theoretical approaches to
    bias analysis of, 226–227
    convergent validity analysis, multicategory variable and polychotomous attribute, 151–153
    validity table for, iterative structural decomposition of, 226
  socioeconomic status and political affiliation, directional hypothesis
    binary class variable and ordinal attribute, 95–97
  stratagraphic rock sections (22) from Wasatch and Uinta Mountains
    structural decomposition of Markov state transition table, 214–215
    structure in Markov transition tables, 189–193
  sympathy and empathy for physicians *versus* undergraduates
    ODA-optimization of Fisher's linear discriminant analysis model, 157–159
    ODA-optimization of logistic regression analysis and interval attributes and hold-out validity sample, 163–165
    ODA-optimization of logistic regression model analysis, 161
  temporal stability of affective experience in 160 undergraduates
    nonlinear reliability, 135–137
    test-retest reliability, 133–135
  temporal stability of learning cycles
    stability of turnover table for, 193–196
  Type A behavior and coronary artery disease
    construct validity analysis of, binary class variable and attribute, 148–149
  Type A behavior and migraine and tension headaches
    discriminant validity analysis, binary class variable and categorical ordinal attribute, 150–152
  Type A behavior and psychological instrumentality
    hold-out analysis of multicategory class variable and interval attribute, 145–148
  Type A *versus* Type B undergraduates discriminated on two ordered personality dimensions

GenODA optimization of multiple training
models, 184–185
GenODA optimization of single training
model and hold-out samples, 182–184
visual acuity of human right and left eyes
multicategory class variable and ordinal
attribute, 112–113
voting behavior of Republicans and, Democrats
in House of Representatives re: Refugee
Act, directional hypothesis
binary class variable and attribute, 90–91
wasps and honey bees, gender of random
samples, nondirectional hypothesis
binary class variable and binary attribute, 88–89
World Series baseball game, time series, lag-1,
serial dependency of sequential baseball
pitches during first two innings
autocorrelation analysis, 200–203
yield (kilograms per 100 square meters) of wheat
(three varieties) as function of fertilizer
(two types)
GenODA analysis from nondirectional random
block design, 180–181
Excel
copying data set from, 48–52
copy and paste method, 48–49
saving as ASCII file method, 50–52
EXCLUDE (EX, EXCL), 31–32
Experimentwise p, 80, 81

F

Fisher's linear discriminant analysis (FLDA)
aging and functional status in ambulatory
medical patients, 159–160
ODA-based optimization of
steps in, 156–157
sympathy and empathy for physicians *versus*
undergraduates, 157–159
Fisher's randomization procedure, for determination
of p for ODA analyses, 77–78
FREE, 32
Fuzzy set theory, 88

G

Generalized ODA. *See* GenODA (generalized
ODA)

GEN (GROUP), variable, 32
GenODA (generalized ODA)
analysis of additive randomized block design,
179–180
for nondirectional randomized block design
yield (kilograms per 100 square meters) of varieties (3) of wheat as function of fertilizer
(two kinds), 180–181
in optimization of multiple training models
Type A *vs.* Type B undergraduates and two
ordered personality dimensions, 184–185
in optimizing multiple suboptimal multiattribute
models, 181–185
Type A *vs.* Type B undergraduates for two
ordered personality dimensions, 182–184
Type A *vs.* Type B undergraduates for two
ordered personality dimensions
LRA classification performance of, 182
LRA training classification performance of, 182
GenODA (generalized ODA) model
in determination of best ODA model for multisample application, 171
in determination of whether single best model
generalizes across multiple samples, 171
directional room-to-vary
patient overall satisfaction, satisfaction with
physician, for different divisions of
medicine, 175
relative generalizability, 177
strict generalizability hypothesis in, 175–176
evaluation across samples, 172–178
fixed-model
fathers' and sons' occupations, from England
and Denmark, and pooled data, 173–174
function of, 170
room-to-vary
patient satisfaction or dissatisfaction with care,
174
relative generalizability in, 174, 175
satisfaction with physician, 174
strict generalizability in, 174–175
selection of
Gen maximum mean PAC heuristic in,
171–172
Gen PAC of Sample, 171
prior odds of samples heuristic, 171–172
as test for absence of paradoxical confounding,
171

GenODA (generalized ODA) model, *continued*
  two-sample application, 170
  weighted, 170
GEN TABLE, 32
GO, 32

H
Hold-out (cross-generalizability) validity. *See also* Validity
  multicategory class variable and interval attribute
    Type A behavior and psychological instrument, 145–147
  multicategory class variable and polychotomous attribute
    fathers' and sons' occupations in England and Denmark, 142–145
    hold-out validity coefficients in, 144–145
Hold-out cross-validation procedure, 69–70
HOLDOUT (HOLD), path/file name, 32

I
ID, variable, 32
INCLUDE (IN, INCL), variable, 32–33
IntegerODA, 236
Interactive standardization, 71
Interpretability, in iterative ODA-based decomposition, 213, 214
Ipsative standardization, 71
Iterative analyses, in conventional statistical procedures, 209
Iterative (sequential) ODA-based decomposition
  stopping rules for
    absolute gain in overall PAC (or standardized effect strength index), 212
    cumulative overall PAC, 213
    interpretability, 213
    number of steps (or models), 213
    parsimony, 213
    relative efficiency, 213
    relative gain in current PAC, 212

L
Leave-one-out (LOO) analysis. *See* LOO (leave-one-out) analysis
Linear model classification
  assumptions in, 237

Logistic regression analysis (LRA)
  involving categorical attributes
    non-optimized, 163
    ODA-based optimization of, physician behavior, personal decision making, and smoking behavior, 163, 164–165
  ODA-based optimization of
    steps in, 156–157
    sympathy and empathy for physicians *versus* undergraduates, 161–162
  use of *vs.* Fisher's linear discriminant analysis, 160–161
LOO (leave-one-out) analysis, 33
  in discriminant validity
    binary class variable and categorical ordinal attribute, 150–151
  in performance assessment, 68–69
LOO (leave-one-out [jackknife]) command, 33

M
Markov model(s)
  chain, 189
  process, 189
  sequential data in
    coded for relative order of occurrence, 187
    creation of transition table for, 180–181
    state transition probability and, 189
    transformation of transition table into transition matrix, 189
Maximum feasible subsystems (Max FS), 11
MCARLO (MC)
  Monte Carlo simulation analysis, 33
  value and, 33–34
Minimum irreducible infeasible subsystems (Min IIS), 11
MIP45 mixed-integer programming formulation
  in general-purpose MultiODA models
    in aggregation of multiple observations of identical attribute values, 232
    to maximize priors-, and/or cost- or return-weighted number of satisfied inequalities, 232
    to normalize discriminant function, 232
MISSING (MISS), variable list, value, 34
Monte Carlo *p*
  confidence levels and, 79
  defined, 79
  simulation research of, 79

Monte Carlo simulation
  applications of, 78
  in statistical aspects of ODA, 79
  in study of statistical methods, 78
Multicategory class variables
  balanced performance heuristic for, 108
  binary attributes and, 108
  comparison with two-category class variables, 107–108
  maximum separation distance heuristic for, 108
  ordinal attributes and
    on perceived waiting time and patient satisfaction with emergency room, 114–115
    visual acuity of human right and left eyes, 112–113
  polychotomous attributes and
    chi-square methods for, 109–110
    congressional voting on 1836 Pinckney Gag rule, 109–111
    directional hypotheses example, 111–112
    log-linear model for, 109
    LOO analysis of classification performance in, 111
    nondirectional hypotheses example, 109–111
    political affiliation status of students and parents, 111–112
    priors-weighted ODA model, 110
    problems with, 109
  a priori selection heuristics for, 107
  sample representativeness heuristic in, 107–108
MultiODA (multivariable ODA), 10–11
  closed-form solutions for, 230
  evaluation of edges and, 231–232
  fast solutions for, 230–231
  future research directions for, 237
  general purpose models
    MIP45 mixed-integer programming formulation in, 232
  hyperdimensional spaces in, 230
  Monte Carlo simulation in, 230
  real world data analysis in, 230
  Warmack–Gonzalez search technique in, 231
MultiODA (multivariable ODA) models
  general-purpose
    in linear case, 232–233
    mathematical programming for, 232
    in nonlinear separating surfaces, 233
  special-purpose
    BooleanODA, 233–234
    ExactODA, 235–236
    IntegerODA, 236
    TauODA, 234–235
    TemplateODA, 236–237
Multiple comparisons
  all possible (exploratory), 81
  alpha splitting in, a priori
    with sequential Sidak procedures, 84–85
    in three-phase study, 84
    in two-phase study, 83–84
  Bonferroni procedure in, 81
  Dunn's Bonferroni-type, 81, 82
  generalized criterion in, 80–81
  planned (confirmatory), 81
  sequential Sidak procedure in, 82
  Sidak per-comparison criterion in, 81–82
Multiple optimal models
  selection heuristics for, 64–67
    balanced performance, 66
    category sensitivity, 66
    distance, 66–67
    maximum mean PAC, 67
    maximum separation, 66–67
    primary, 67
    prior odds, 67
    random selection, 67
    sample representativeness, 65–66
    secondary, 67
Multiple sample analysis
  for between-sample differences, 167
  for between-sample similarities, 167
  determination of best ODA model for, 168
  fixed ODA and, 167
  generalizability algorithm in, 168
  ODA generalizability algorithm, 168, 170–172
  room-to-vary ODA and, 167
  sample pooling and Simpson's paradox, 168–169
Multivariable models
  multiattribute multivariable ODA, 155
    nonlinear, 155
  optimizing (See Optimizing suboptimal multivariable models)
  single attribute univariable ODA, 155, 156
  suboptimal
    Fisher's linear discriminant analysis, 156
    linear regression analysis, 156
    Y-hat, single score observation, 156
Multivariable ODA. See MultiODA (multivariable ODA)

N

Nonlinear reliability models, types of, in ODA paradigm, 137–138
Normative standardization, 70–71

O

Observation, classification of, 9–10
Observed score
   in classical test theory, 223
   in ODA, 223
ODA (optimal data analysis)
   definition of predictor, 7–8
   definition of predictor variables, 8
   definitions of, 9
   historical perspective, 11–12
   hypothetical applications of, 11–27
     astrology, 11–12
     astronomy, 12
     beer brewing, 12–13
     bird watching, 13
     credit collection, 13–14
     credit screening, 14
     criminal justice, 14–15
     dating, 15
     direct mail advertising, 15–16
     driver licensing, 16
     epidemiology of AIDS, 16–17
     farming, 17
     fishing, 17–18
     gambling, 18
     golfing, 18–19
     history, 19
     hostage negotiation, 19–20
     hurricane forecasting, 20
     life insurance, 20–21
     missionary work, 21
     personnel selection, 21–22
     prospecting, 22–23
     selling, 23
     speeding tickets, 23
     suicide, 24
     target recognition, 24–25
     teaching, 25–26
     vacationing, 26
     weight loss, 26–27
     zoology, 27
   mathematics and statistics capabilities in, 229
   specification of weights, 8–9
   superiority of
     for conceptual clarity, 5, 6
     for ease of interpretation, 5
     for maximum accuracy, 5, 6
     for valid Type I error, 5
   users of
     community of scientists, 240
     statistically and computationally advanced, 239–240
     statistically and computationally restrained, 240
   vs. maximum likelihood paradigm, 4
   vs. ordinary least squares paradigm, 3, 4
ODA (optimal data analysis) generalizability algorithm. *See* GenODA (generalized ODA)
ODA (optimal data analysis) model
   classification error rate of, 69–70
   definition of, 64
   obtaining of
     cutpoints in, 63–64
     effect strength for sensitivity in, 64
     hypothetical case, 61–62
     sensitivity in, 64
   stability assessment of, 68–70
ODA (optimal data analysis) software
   advantages of, 5
   audience for, 6
   commands in, 29–37
   ease of learning, 5
   ease of teaching, 5
   ease of use, 5, 6
   MS-DOS command prompt window, 37
   Programmer's File Editor, 37
One-sample jackknife, in performance assessment, 68
OPEN, path/file name/DATA, 34
Optimal cutting score (OCS), 10
Optimal data analysis (ODA). *See* ODA (optimal data analysis)
Optimization
   ODA-based
     of Fisher's linear discriminant analysis, 156–157
     of logistic regression analysis, 156–157
     research in, 157
Optimizing suboptimal multivariable models. *See also named method, e.g.,* Fisher's linear discriminant analysis (FLDA)
   complex models, 165

Fisher's linear discriminant analysis, 157–160
    linear regression analysis, 160–165
Ordered attributes, 8
Ordinal attributes
    directional hypothesis, binary class variable and ordinal attribute
        relative strength of gender *vs.* other effects, 100–101
        socioeconomic status and political affiliation, 95–97
    multicategory class variable and
        on perceived waiting time and patient satisfaction with emergency room, 114–115
        on visual acuity of human right and left eyes, 112–113
    nondirectional hypothesis, binary class variable and ordinal attribute
        efficiency of two auditing methods, 94–95
    nondirectional hypothesis, binary class variable and ordinal categorical attribute
        outcome of marital therapy, 97–98
        outcome of two headache therapies, 98–99
Ordinal data
    log-linear model and, 93, 94
    Mann-Whitney *U* test and, 93–94
Ordinal scales
    categorical, 93
    Likert-like, 93
OUTPUT, 34

P

p
    empirical value of, 80
    experimentwise, 80, 81
    Fisher's randomization methodology for ODA analyses, 77–78
    Monte Carlo, 79
    for ODA analysis, analytic method, in non-weighted applications, 74
    rarity *versus* robustness of, 74
    target value of, 80
PAC (percentage accuracy in classification), 9
    absolute gain in overall, 212
    cumulative overall, 213
    current, relative gain in, 212
    definition of, 57
    overall 58

Parsimony, in iterative ODA-based decomposition, 213
Percentage accuracy in classification. *See* PAC (percentage accuracy in classification)
Permutation probabilities, Fisher's randomization procedure and, 77
Permutation probability method, 75
PFE (programmer's file editor), opening ASCII file in, 52
Polychotomous attributes
    analysis of, dummy-coded binary attributes in, 91
    nondirectional hypothesis, binary class variable and polychotomous attribute
        gender and rheumatic disease data, 91–93
Practical significance, 73–74
    drift and, 74
Predictive value. *See* PV (predictive value)
PRIMARY (PRI), criterion for multiple optimal solutions, 34–35
PRIORS, 35
Programmer's File Editor (PFE), 37, 38
    Change to File's Directory in, 40
    Command Output in, 43
    data files, 39
    data set creation with, 45–47
    directory, 38
    examples, 39
    execution of analysis, 41–43
    repeat analysis in, 44
    revision of script file in, 44
    script file creation with, 45–47
    scripts, 39
PV (predictive value)
    base rates and, 60
    defined, 57, 59–60

Q

Qualitative attributes, 8
QUIT, 35

R

Randomized block design (RBD)
    ANOVA for, 178–179
    GenODA for, 179–180
Regression, in nonlinear reliability models, 137–138
Relative efficiency, in iterative ODA-based decomposition, 213

Reliability analysis
    bias and random error and, 224–225
    inter-rater
        in cardiologists (two) evaluating 200 electrocardiograms, 122
        chi-square approach to, 123–124
        correlation coefficient r, 124
        intraclass correlation coefficient of reliability, 124
        kappa coefficient for nominal data, 124
        multivariate, 122
        ODA paradigm, 125
        ordinal attribute in, cardiologists' rating of electrocardiograms, 125–126
        ordinal attribute in, efficacy study of neuroleptic maintenance dosage, 126–127
        percentage accuracy in classification in, 123
    intraclass correlation in, 138–140
        assumptions in, 138
        dentists' (four) ratings of state of ten patients' teeth, 139–140
        fixed effect reliability study in, 138–139
    iterative structural decomposition in
        for bias or random error, 223–225
    nonlinear
        degenerate solution, 137, 138
        local regression throughout range, 138
        reliable at extreme values; regression at intermediate, 137–138
        stability of attribute at higher values; regression at lower, 137–138
        temporal pattern of undergraduates' emotional experience, 136–137
    parallel forms
        in classical test theory, 128
        correspondence between two procedures for assessing Type A behavior, 128–130
        equivalence coefficient in, 128
        ordered attributes, 128–130
        Pearson product-moment correlation coefficient in, 128
    reliability coefficient in, 121–122
    split-half
        assessment of psychological androgny in male undergraduates, 130–132
        polychotomous attribute in assessment of psychological androgyny, 130–132
    temporal, 133
        reactivity bias in, 132
        stability of undergraduates' emotional experience, 133–135
Reliability table
    inter-rater, iterative structural decomposition of cardiologists (two) classifying 200 electrocardiograms into three diagnoses, 224–225
    iterative decomposition of, 223
    iterative structural decomposition of
        ODA nondirectional bias analysis of, 224–225
Repeated measures (within-subjects) analysis
    comparison of responses of single group of observations to two or more different attributes assessed at a single testing, 206–207
    comparison of single group of observations to single attribute assessed across two or more testings, 204–206
    prospective study of efficacy of lecture- and case-study-based methods in emergency medicine, 205–207
    generalizability of directional alternative hypothesis, 206
    ODA model, 205–206
    prospective study of efficacy of lecture- and case-study-based methods in emergency medicine residency
        effect generalized across sex, 206
        gender discrimination based on score, 205, 206
        increase in score over course of training, 205, 206
        ODA model, 205–207
    single-factor within-subjects design, 204–205
        conventional analysis, 204–205
Replication validity
    in performance assessment, 69–70
REPORT (REP), 35
RESET, 35

S

SAS, creating ASCII files from, commands for, 54–55
Saving Excel data set
    as ASCII file, 50–52
        comma-delimited, 51
        tab-delimited, 50
Script file, creation with PFE, 45–47
SECONDARY (SEC), 35–36

SEED, value and, 36
Sensitivity
　defined, 57, 58
　effect strength for, 59, 64
　weighting by prior odds and, 58
Sequential analysis. *See also named type, e.g.* Autocorrelation (time-series) analysis
　autocorrelation (time-series), 198–203
　in identification of structure of Markov transition tables, 187–193
　repeated measures (within-subjects), 204–207
　single-case (N-of-1), 207–208
　　classical test theory for ten or fewer points, 207–208
　　ODA for measurements over ten points, 208
　　parallel use of ODA in, 208
　of turnover tables, 193–198
Shuffle (permutation), in Fisher's method, 77
Sidak procedures
　nonsequential, 85, 86
　sequential, 82
　　with alpha splitting, 84–85
　　in multiple comparisons, 82, 84–84, 86
Simpson's paradox, 168–169
　for applications involving only categorical data, 168–169
　for applications involving only ordered data, 169
　　normative standardization of X and Y, 169
　　verification of X and Y homogeneity across samples, 169–170
　remedies for, 169
SPSS data set
　creating ASCII files from, 52–54
　　save as Fixed ASCII (*.dat), 52–53
　　Save Data As tab-delimited (*.dat), 54
　　selection of variables for, 53
Stability
　assessment of
　　bootstrap methods, 69
　　hold-out cross-validation procedure, 69–70
　　leave-one-out validity analysis, 68–69
　　one-sample jackknife in, 68
Statistical significance
　analytic methodology and, 74–76
　description of, 74
　Fisher's randomization methodology, 77–78
　p greater than 0.05, 74
　vs. practical significance, 73

Structural decomposition with sequential data
　people completing six consecutive monthly interviews re: voting intentions in 1940 presidential election, 220–223
　stability and change in turnover tables, 220–223
　strategraphic rock sections (22) from Wasatch and Uinta Mountains
　　Markov state transition table, 214–215
　temporal stability of learning styles
　　analysis to determine statistically reliable sequential structure for off-diagonal elements, 216–217
　turnover table for exchange of material and psychological resources, 217–220
Student's t test
　ODA *vs.*, 76

T
TABLE (FREE TABLE), row/col, 3636
target value, of p, 80
TauODA
　in analysis of data in ordered categories, 234
　in linear regression, metric-free, 234–235
　　investigation with Monte Carlo experiments, 235
　in maximizing goodness-of-fit between actual and predicted category assignments, 234–235
TemplateODA
　in design of optimal templates, 236–237
　formulation as pure integer problem, 237
Temporal stability of learning cycles
　ODA-model for
　　exploratory analysis for sequential structure of off-diagonal elements, 216–217
　　stability of turnover table for, 193–196
TITLE, 36
Total effect strength, defined, 61
Transformations
　standardization of
　　interactive, 71
　　ipsative, 71
　　normative, 70–71
Transition tables
　ODA in identification of structure underlying types of rocks within carbonate units, 189

Transition tables, *continued*
  assignment rules in, 191
  nondirectional categorical ODA, 190–191
  one-dimensional sequentially ordered structure in state transition table, 192
  results, 193
  starting heuristic for selection of assignment rule, 191–192
Turnover tables
  assessment of stability and instability in, 194
  change assessment in
    nondirectional analysis, 221
  consecutive codes nonrepeatable
    ODA directional nonlinear hypothesis for, 217-219
  function of, 193–194
  ODA assessment of stability in exchange of material and psychological resources, 196–198
    dissimilarity hypothesis in, 196–197
  ODA assessment of stability in temporal stability in exchange of material and psychological resources
    consecutive codes not repeatable, 196
    in exchange of material and psychological resources, 196
  ODA assessment of temporal stability of learning, 194–196
    classification of observation type in, 194–195
    directional ODA model in, 195–196
  ODA model of
    final GenODA model, 221, 222
  problems in conventional analysis of, 194
  similarity hypothesis for
    clockwise hypothesis, 219–210
    counterclockwise hypothesis, 219
  stability assessment in, 220–221
Type I error rate
  multiple comparison in, 81–83

per comparison *vs.* per experiment, 80
a priori alpha splitting in, 83–86

U
UniODA (univariable ODA)
  capability of, 230
  defined, 10

V
Validity
  as appropriateness of label attached to measure, 141
  in classical test theory, 225
  construct (*See* Construct validity)
  convergent, 149, 150 (*See also* Convergent validity)
  discriminant, 149, 150 (*See also* Discriminant validity)
  hold-out (cross generalizability) (*See* Hold-out (cross generalizability) validity)
  as legitimacy of estimated classification performance, 141
Validity table, convergent, iterative structural decomposition of, 225–227
Value of the likelihood function, ML paradigm and, 4
Variance ratio, OLS and, 3
VARS, variable list, 36–37

W
Weighting, by prior odds, 58
WEIGHT (RETURN), variable and, 37
Weights
  prior odds, 9
  quantitative assessment of value/importance to attribute, 9
  specification of, 8–9

About the Authors

Paul R. Yarnold received his PhD in academic social psychology from the University of Illinois at Chicago. He is currently a research professor of emergency medicine at the Feinberg Medical School of Northwestern University and an adjunct professor of psychology at the University of Illinois at Chicago. He is a fellow of the Society of Behavioral Medicine; fellow of Division 5 (Measurement, Evaluation, and Statistics) and 38 (Health Psychology) of the American Psychological Association; sits on the editorial boards of *Perceptual and Motor Skills* and *Educational and Psychological Measurement*; and has authored approximately 200 articles in the areas of medicine, psychology, and statistics. He is coeditor of *Reading and Understanding Multivariate Statistics* and *Reading and Understanding More Multivariate Statistics*.

Robert C. Soltysik received his MS in industrial and systems engineering from the University of Illinois at Chicago in 1983. He is currently a scientific programmer for the Veterans Affairs Chicago Health Care System. He has consulted for many large companies in the fields of operations research and statistics, including the development of a press sequencing optimization system for magazine manufacturing, a personnel-selection system for a national drugstore chain, and interactive Internet-based clinical intervention systems. He is the codiscoverer of the ODA paradigm, has created ODA software, and has authored nearly two dozen articles concerning ODA.